In the Company of Crows and Ravens

Yale University Press   *New Haven and London*

Posturing crows

In the Company of

Crows and Ravens

John M. Marzluff and Tony Angell

Illustrated by Tony Angell   Foreword by Paul Erhlich

"Dust of Snow" from *The Poetry of Robert Frost,* edited by Edward
Connery Lathem. Copyright 1923, 1969 by Henry Holt and Company,
copyright 1951 by Robert Frost. Reprinted by permission
of Henry Holt and Company, LLC.

Excerpt from Ted Hughes, "Crow and the Birds," reprinted
by permission of Faber and Faber, Ltd., London

Designed by Rebecca Gibb
Set in Bulmer type by BW&A Books, Inc.
Printed in the United States of America
by Edwards Brothers, Ann Arbor, Michigan.

The Library of Congress has cataloged the hardcover edition as follows:
Marzluff, John M.
In the company of crows and ravens / John M. Marzluff and
Tony Angell ; illustrated by Tony Angell ; foreword by Paul Ehrlich.
p. cm.
Includes bibliographical references (p.   ) and index.
ISBN 0-300-10076-0 (clothbound : alk. paper) 1. Crows. 2. Ravens.
3. Human-animal relationships. I. Angell, Tony. II. Title.
QL696.P2367M358   2005
598.8'64—dc22   2005007008

A catalogue record for this book is available from the British Library.

The paper in this book meets the guidelines for permanence and
durability of the Committee on Production Guidelines for Book
Longevity of the Council on Library Resources.
ISBN 978-0-300-12255-8 (pbk. : alk. paper)
10 9 8 7 6 5 4 3 2 1

The way a crow

Shook down on me

The dust of snow

From a hemlock tree

Has given my heart

A change of mood

And saved some part

Of a day I had rued

Robert Frost, "Dust of Snow," 1923

Contents

Foreword

In a world of disappearing biodiversity, enormous attention has been paid
to the impacts on birds of growing numbers of people and accelerating
consumption. From the Dodo to the Ivory-billed Woodpecker to, now, Indian vultures, people have been concerned with the extermination of beautiful and interesting creatures, many of which play key roles in supplying vital ecosystem services to humanity. But relatively little attention has been
paid to the other side of the coin, the manifold influences some groups of
birds have had on human culture. *In the Company of Crows and Ravens*
helps to redress the balance—and in the process gives us a splendid overview of this group of intelligent birds.

Members of the crow family are, perhaps second only to House Sparrows and the pigeon, the most prominent features of urban landscapes in
much of the world. And they are infinitely more interesting than most of

our other avian close companions. They are much favored by human activities. We create the roadkills and garbage that they love to devour. They are even likely to snatch your fast-food hamburger from the picnic table, undeterred by its lethal fat content. And human suppression of many of the raptors that would devour them makes cities and agricultural areas relatively safe havens, especially since they are more than smart enough to outwit people with shotguns. Crows, ravens, and Rooks, of course, are themselves vigorous predators and competitors—by helping them prosper, we heighten the jeopardy of more specialized bird species.

Not only do corvids tend to thrive in human-made environments, but they have also penetrated our psyches. Scarecrows have become characters in fiction and "Maître Corbeau" teaches us lessons in proper behavior. We "eat crow," climb to the "crow's nest," tear things apart with a "crowbar," and are "ravenous" after a long hike without food. Many human groups have considered them gods. The high intelligence and humanoid behavior of these birds have long fascinated biologists, and have resulted in some wonderful accounts of the comportment of individual species by, among others, Professor Bernd Heinrich of the University of Vermont. But this book is the first to produce a grand overview of the human-corvid complex and is truly something to crow about.

With any luck *In the Company of Crows and Ravens* will stimulate more people to connect with nature by enjoying the antics of these easily observed birds, and learning about them and about the complexities of nature in the process. If people can only learn to appreciate biodiversity in the process, then they will join the growing numbers who look at the paving over and plowing under of natural habitats and quote the raven—"nevermore."

Paul Ehrlich

Preface

It's midmorning, and a single crow has arrived unseen to parade about the lawn outside our open window. Failing to arouse us with visual cues, the crow leaps to the outside railing and calls: two loud, sharp caws in succession. Alerted, we look up from our work, and one of us walks obediently to the kitchen to retrieve a previously saved crust of bread. This morning ritual of bread-fetching for our saucy friend has been going on long enough to become habit.

A subtle transaction has occurred between two species that routinely find themselves sharing common ground. A crow has influenced what we do, and we have affected its behavior. Leaving aside the possibility that the crow has derived some twisted satisfaction in first startling and then manipulating us, the bird has certainly benefited from the delivered meal, and we have been rewarded by an intimate moment with another species, not to mention the peace and quiet that ensued once the crow was fed. A good argument can be made that we have trained each other to behave in a way that is mutually beneficial. At first such an esoteric benefit to a person may seem insignificant, but even brief encounters with crows can be powerful. Robert Frost commemorated his crow moment in his poem "Dust of Snow," which we reproduce as our epigraph to this volume.

Let your mind wander a bit now and consider our language, art, culture, and religion. They have all been influenced over the ages by our interactions with crows. Likewise, crows have benefited enormously by associating with people. We have reduced many of their natural predators and issued royal, political, and religious edicts that have forbidden the killing of some crows. Many of the various species known generally as "crows" are provisioned on massive scales by our agricultural and urban wastes. Is this a uniquely strong and positive relationship between people and a wild animal? We think it is.

Crows and people are engaged in a type of mutualism somewhat akin to the one between shark and remora or clownfish and sea anemone that you may have learned about on popular television nature shows. Just like the shark that provides its gills, mouth, and other body parts as foraging sites for a remora in exchange for parasite removal, we, by choice or circumstance, provide living space and food for crows in exchange for intellectual, emotional, and aesthetic stimulation. We may also reap direct benefits as crows clean our environments of wastes and pests, and by observing their behavior and noting their mortality, we are warned of harmful pathogens like rabies and West Nile virus. We wrote this book to illustrate and document this apparent mutualism, as well as other positive and negative interactions between crows and people. Our suggestion that a portion of the relationship represents a novel evolutionary force, one we term *cultural co-*

A wild crow at the authors' window waiting for food

evolution, is a new hypothesis we introduce in the first chapters and explore throughout the book.

Our fascination with crows runs deep. But are we unusual for our species? Perhaps we are rare in the extent to which we engage our black-feathered neighbors, but our research suggests that strong interactions between crows and people are anything but unusual. We are convinced that crows and people share a long and storied history—a history that probably started among ancestral crows and our primate ancestors and that has subtly molded both crows and people.

Is this history so obvious that it requires no elaboration? Many would question the need for a whole book about crows. Arthur Cleveland Bent, one of the leading chroniclers of bird lore in the twentieth century, wrote: "If a person knows only three birds one of them will be the crow." What he really meant, though, was that everyone *recognizes* the crow. To us this is not the same as *knowing* the crow. Our neighbors often remark that they can tell us everything we need to know about crows in just a few words. They are "noisy," "destructive," "dirty," "aggressive," and, yes, "clever"! This epitomizes many people's views of crows as pests, elusive targets for shooters, consumers of songbirds, garden raiders, and symbols of death and filth. And yet others have and do revere crows as spiritual totems, admire their sagacity, and find amusement in their playful and problem-solving antics. In our experience, few people hold no opinion about crows. Crows' conspicuous, bold, and brash ways make most people take notice and form opinions. Differing points of view result because we all know, or respond most strongly to, only a part of the crow. Here we attempt to meld the various facets of the crow so that you can indeed know the whole animal.

Are the various impressions about crows simply a reflection of the variety of crow species around the world, or are all "crows" viewed similarly? Our research suggests that most of the forty-six species generally known as "crows," including actual crows, Rooks, jackdaws, and ravens, are viewed similarly by most people. Many have been both loved and hated through history. Contrasting their ecologies in interactions with people allows us to weave a richer story. For clarity, however, we use lowercase to refer to general and inclusive groups of species like crows and ravens but capitalize names when we refer to a specific species, such as the Common Raven or American Crow.

Crows and people interact in a variety of influential ways. We detail four basic relationships between crows and people in the chapters that follow. We open with a close look at the influences people have on crow evolution and ecology. Next we examine what crows have done to influence human cultures throughout the world. We follow this with a detailed look at the cultural life of crows, exploring their behavior and traditions and our influences on them. Last, we open a discussion of what crows seem to be doing to affect our ecology and evolution, and we pose some of the still-unanswered questions about crows that make them such an intriguing subject of study. We include an appendix with suggestions on how you can help answer some of these questions inasmuch as this demonstration of mutual influence is just beginning to be understood and it is quite clearly an ongoing and evolving phenomenon. In summarizing a unique history of association, we make the case for cultural coevolution between crows and people. Along the way we reveal the diverse culture, behavior, intelligence, and adaptability of crows and depict their elegance and ecological role.

Our transition from a historical association with ravens to a contemporary world where crows are our main associates provides the second organizational thread to our story. We first review interactions among people and crows on many continents, emphasizing how these species and people influenced one another. As we delve deeper into the relationship between crows and people, we focus more narrowly on American Crows (*Corvus brachyrhynchos*) and modern North Americans. Through this single common species we can explore and understand many of the social strategies, physical attributes, and examples of intelligence that have contributed to this bird's extraordinary success amid a world of change. Along the way, we discover that crows and people share similar traits and social strategies. To a surprising extent, to know the crow is to know ourselves.

As we think about people's interactions with crows, we must ask some difficult, soul-searching questions that may refine our attitudes and values toward other animals. Because crows have been so successful at exploiting our excesses, some people view them as serious pests. Increasingly, crow populations are being controlled through lethal means, which forces society to argue about the ethics of killing crows and other cognitive creatures. The resolution of such arguments may ultimately help form an ethic that greatly affects our cultural trajectory. Will we be quick to pull the trigger or will we think about how best to balance our effects on the ecosystem? Are we conquerors or stewards of nature? Knowing the crow argues for stewardship, if only for a selfish reason. Simply stated, to conquer the crow is to stifle one of our powerful and long-standing cultural motivators. Our art, religion, literature, language, and music all draw on crows for inspiration. Life without such inspiration would be less creative, less imaginative, and less mysterious. Our investigations of crows and people have convinced us that people need crows specifically, and nature generally, if we are to maintain a vibrant culture. After reading our story, we hope you agree.

Acknowledgments

Marzluff's students, postdoctoral fellows, and colleagues helped us view the crow's world. Rick Knight and David Hutchinson kept our early focus on crows. Marco Restani introduced us to ravens in Greenland. Gier Sonerud, Dieter Glandt, and Michael Abbs helped us understand European crows. Erik Neatherlin, John Luginbuhl, and Jeff Bradley showed us the canopy and exposed the lives of rainforest crows and ravens. Scott Derrickson, Peter Harrity, Laura Landon, and Kate Whitmore taught us about reintroduction. Stacey Vigallon helped us see nests from a jay's eye. Bob Reineke, John Withey, Bill Webb, Kara Whittaker, Thomas Unfield, and Stan Rullman read the entire manuscript, checking our science. Ellen West scrutinized the manuscript and tracked down leads to literature in cyberspace and numerous libraries. Sharon and Vince Penn told us about Quileute ways. Gene Hunn, Scott Derrickson, Tino Aguon, Rob Fleischer, Rob-

Acknowledgments

ert Askins, Reiko Kurasawa, Hito Higuchi, Cynthia Sims Parr, Ed Murphy, Steve Emslie, Uli Reyer, Hussein Isack, Richard Connor, and Jay Nelson checked our reasoning, updated our information, and made their unpublished research freely available. Rene Drake, Kevin Grigsby, Bob Ricci, Reid Hargis, Beth Wapelhorst, Lee Bond, Ed Bessetti, Dave Manuwal, and Gordon Bradley shared with us their personal encounters with crows. Linn Catlett, Dave Peterson, Ivan Eastin, Jim Agee, Steve West, Colleen Marzluff, Tina Blewett, Fred Lohrer, Stuart Houston, Jeanne Shepard, and Paul Boardman pointed us to fascinating literature and always encouraged our curiosity. Eric Shulenberger, Burt Lewis, Robert Askins, and two anonymous reviewers checked our prose and reasoning.

Eric Sorensen led us through journalism boot camp and graciously edited the entire manuscript. He inspired us with his writing and interest in crows and ravens. We will never look at *but*s, parentheses, and tri-syllabic words in the same way again, thanks to Eric.

Jean Thomson Black encouraged, edited, and broadened our appreciation for crows. Along with her colleagues at Yale University Press, Laura Jones Dooley, Molly Egland, Laura Davulis, and Heidi Downey, she painlessly and efficiently clarified our thoughts and improved our writing.

Russ Balda and Bernd Heinrich kindled, encouraged, directed, focused, and honed our investigations on corvids. They have always been eager to listen to, critique, and discuss our ideas. Thanks, too, to Boria Sax, Lynne Grinstein, Bill Holm, Les Perhacs, and Thomas Quinn, who opened doors and shared historical insights and field notes that gave greater authenticity and vitality to our narrative and drawings. Ivan and Carol Doig provided their usual high standard of narrative and illustrative expectation and encouragement that helped us maintain the necessary stamina to complete what we set out to do.

Acknowledgments

Much of our research was supported by the National Science Foundation, University of Washington, Olympic National Park, Olympic Natural Resources Center, Pacific Northwest Forest Sciences Laboratory, Rayonier, United States Fish and Wildlife Service, Sustainable Ecosystems Institute, and Washington State Department of Natural Resources.

Colleen, Lee, Zoe, Danika, Gavia, Larka, Bryony, and Gilia, our wives and children, have supported us at every step. They ate crow, chased crow, watched crow, caught crow, drew crow, enjoyed crow, read crow, and collected crow. But never tired of crow. They excused our numerous absences and distracted ways during the years of writing.

We thank you all for your time, interest, and inspiration. May you see crows in groups of twelve directly above you at sunset.

Overleaf: Sighting twelve crows at sunset foretells good fortune

ONE

Cultural Connections

Crows demand our attention. When they fly, we two watch and follow. When they call, we listen. When they encounter people, we antici-pate fascinating interactions. We are ornithologists and artists consumed by these common black, noisy birds. Thousands of days spent chasing, draw-ing, watching, sculpting, searching for, and reading about these birds have rewarded us with a deep understanding of their biology and an unashamed respect for their abilities to frustrate, challenge, inspire, and exploit us. Few people are as passionate about birds generally, or crows specifically, as we are. But our scientific and artistic passion for crows did not prepare us for the impact these birds would soon have on our basic thoughts.

Yes, we knew that crows were flexible and clever problem solvers with complex social lives, culture, and communication abilities. But their connec-

Cave art from Lascaux, France, showing a hunter's death. The bird-headed man is thought to represent the external soul of the prehistoric hunter. The bird head and the nearby carved bird on a stick are likely crows or ravens because of their mysterious yet conspicuous association with death.

tions to, and influences on, humans are far stronger than we imagined. We discovered that crows do not affect just us. They have profoundly affected people all over our planet at every step of human evolution. Cave dwellers scratched images of crows on their walls. Early hunters and gatherers built scarecrows to keep crows from their drying fish and carved totem poles in their honor. Noah counted on crows to find land. Hindus and Japanese sought their wisdom. Shakespeare wove them into his plays. Franz Schubert commandeered their melancholy way in his song-cycle *Die Winterreise*. Alexander the Great mistakenly ignored a dying flock of crows as he stood before the gates of Babylon. His soothsayers warned him that evil would follow, and within two weeks he was dead. Weeks before taking his own life, Vincent Van Gogh painted crows. Rock bands name themselves after them. Hundreds of thousands of ordinary people have desperately sought to understand them. Crows have power—an unusually potent cultural staying power.[1]

There are more than forty distinct species of "crows," technically members of the genus *Corvus,* worldwide, which people interact with to various degrees. These include primarily meat-eating "ravens" and omnivorous "crows, rooks, and jackdaws" that favor fruit, seeds, and insects. Familiar jays and magpies, nutcrackers, and obscure choughs are close relatives to crows that round out the biological family collectively known as corvids. We describe and distinguish among many species in the next chapter.

Over time people have come to interact less with ravens and more with crows. This seems due in part to expanding populations of most crow species, whereas ravens have adapted less well to human presence, and may well be discouraged by aggressive crow mobs. Important interactions with crows and ravens probably began when our nomadic ancestors hunted and fished. Indeed, the earliest expressions of humankind recorded thirty thousand years ago in the caves of Lascaux, in modern-day France, feature a crow-headed man and a totemic bird form thought to represent the external soul. We imagine that it didn't take an enterprising Common Raven (*Corvus corax*) long to associate the triumphant early European hunter or fisher with the possibility of sneaking in to take its share of the carnage. Maybe we shooed away this thief; more likely, we offered food in homage to a kindred spirit associated with the successful kill. Nomadic hunters surely took note of the raven's distinctive calls, savvy, and persistence. Similar interactions must have been common wherever ravens ranged. Aboriginal Australians, for example, were flanked by Australian Ravens (*Corvus coronoides*), which still scavenge around towns and camps to this day.[2]

Some evidence of the raven's influence remains. The Hän people of the Yukon, in Canada, for example, mimicked the raven's calls to attract bears to their hunting areas. Likewise, Eskimo hunters in Greenland associated ravens passing overhead with nearby caribou. Over time, the nearly constant contact among hunters, fishers, and ravens affected our early folklore and spiritual beliefs. Crows and ravens were integral to Tibetan funeral rituals until the 1950s: a dead loved one was ceremoniously cut into small pieces and placed on an altar; crows, ravens, and other scavengers then carried the departed to the next life. Eskimos tied the foot of a raven around their newborn babies' necks so that as adults they would be able to endure long periods without food. Indigenous peoples of North America viewed the raven as a creator, trickster, and messenger. Clans were formed in the bird's name, and its dramatic shape and demeanor inspired carved totemic images, myths, dances, and song.[3]

Europeans, too, formed a mythic relationship with ravens. The Nordic god Odin learned about the world from ravens. His corvid connection is evident today in Stockholm, where the Hotel Oden is adorned with raven art, and in Oslo, where a woodcarving outside the city hall depicts Odin in the company of ravens. Good luck was expected by the Irishman who saw a raven croaking as it flew to his right side. Scottish Highlanders stalking deer considered a calling raven a sign of good luck. Heraldic figures of ravens and crows were often chosen to represent medieval families and clans.[4]

Over time, people roamed less, settling down to farm. In turn, we started interacting more with Rooks (*Corvus frugilegus*), Western Jackdaws (*Corvus monedula*), and various other species of crows. They shared our crops and cities, and benefited from the conversion of forests to fields. Crows seemed especially suited to capture our ancestors' imaginations. On the Faeroe Islands, in the northern Atlantic, an unwed girl would throw a stone, a bone, and a piece of turf in quick succession at a Hooded Crow (*Corvus corone*) to find out who her husband would be. If the crow flew toward the sea, her husband would arrive from the sea. If the crow flew toward a town, then he would come from the town. But if the crow did not fly, the girl would not marry. The ancient Greeks often associated crows with the god Apollo. Apollo's unfaithful mistress, Coronis, is the source of the word "corone," Greek for "crow" and the modern scientific name for the Hooded Crow. Greek culture mirrored Native American culture by portraying crow as a clever liar who was eventually banished to the sky. The crow can still be seen as the constellation Corvus in the night sky riding on the back of Hydra, the water snake. Roosting crows signaled the start of the Sabbath to early Hebrews.

The evolving agriculturally based human civilizations wove, painted, and carved crows into impressive works of art. Their language, too, reflected the bird's constant presence. A leader's nickname might pay homage to the crow. Viking Chief Olav Tryggvason was called "Crowbone" because of his uncanny ability to predict the future from bird bones. If a nun was deemed "bad" by French peasants, she was known as a crow. Specific omens were associated with the number, timing, and direction of crow sightings. Even subtle differences in the encounter conveyed profoundly different messages. In some eastern mythologies, for example, finding a dead crow in the road was good luck, but having a live one pass in front of you was bad.[5]

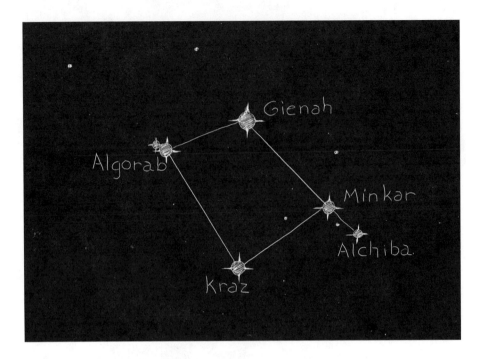

The constellation *Corvus,* the crow

As civilizations expanded, fostering conflict and the spread of disease, crows and ravens prospered, but their image began to suffer. The birds feasted on the corpses that littered medieval battlegrounds. People, however, interpreted this predictable biological response as a supernatural sign and came to view crows and ravens as omens of bad luck and harbingers of death. A flock of ravens was called an "unkindness," a group of crows a "murder." When the Black Death swept across western Europe, physicians tending the dying were wrapped in waxed robes and a helmet that mimicked the crow's likeness. The doctors were to be avoided. Even today this negative association is strong, often stirring our fascination with mystery, death, the occult, and horror. Hordes of crows starred in Alfred Hitchcock's movie *The Birds,* and a single raven inspired Edgar Allan Poe's famous poem "The Raven." Some modern Japanese still call on wildlife managers to move nearby crow roosts because of the belief that their presence foretells death.[6]

The rise of the modern city brought us into even greater contact with crows. Crows flourish with the milder climates, abundant food, and protection from harassment that our cities provide. Their familiarity has enriched the culture of marketing (think of Three Crow Brand spices or Crows candies), street people (some homeless call themselves the "Tribe of Crow"), and language ("as the crow flies" or "crow's nest," for example). The Black Crowes and Counting Crows are popular modern rock groups. Challenges posed by crows also affect our culture through industry (crop protection devices, or *scarecrows*) and art (Van Gogh's final painting was likely *Wheat Field with Crows*). Mark Twain, although enamored with crows during his travels around the world, feared what might become of them in India, where he quipped that they "cost the country more than the government!" Crows continue to act as barometers of our quality of life in our largest modern cities. In Tokyo, exploding crow populations point to a need for better garbage removal. In New York and many other

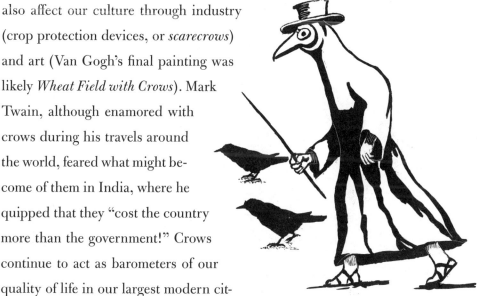

During outbreaks of the Black Death, medical doctors wore helmets reminiscent of crow heads. Perfume was held in the large "beak" to reduce the stench of rotting corpses.

The intersection of crows and people through time. A mutually reinforcing relationship is seen through the ages. Ravens scavenged from large animal kills in the Pleistocene and quickly learned to exploit the foods gathered by early human hunters and fishers. Fending off scavengers may have favored people with a culture of living in groups. The mysterious nature of crows and ravens made them likely subjects for worship and mythology from one to thirty thousand years ago, illustrated here by Northwest Indian cultural icons and the Scandinavian myth of Odin, to whom two ravens spoke daily. Counterparts in Asian cultures were often celebrated in carved and painted forms, shown here by a large, seventeenth-century Edo mural of a crow gathering. The battles and wars that raged from the Middle Ages to the late nineteenth century were extraordinary feeding grounds that increased people's contact with scavenging crows and ravens. This and competition with an increasingly agrarian civilization led to negative attitudes toward the species. Medieval plague doctors wore corvid-like helmets. Crows were hunted. Despite this, crows and ravens have been incorporated into our culture since the nineteenth century through literature, art, film, music, and sports. Modern scientists study the response of crows in our ever-more urban world. Today, crows thrive in suburbs but increasingly die there from new challenges like West Nile virus. The intersecting path continues.

American cities, dead crows are alerting concerned citizens and governmental agencies to the hazards of a rapidly spreading exotic disease, West Nile virus. Crows even leave their mark on modern recreation. School and professional teams emblazon their uniforms with crow and raven symbols and have corvid mascots. Campers in nature parks

A box of Three Crow Brand spice from Rockland, Maine

are discouraged from feeding wildlife, partly in an effort to stem the colonization of isolated reserves by otherwise rare crows.[7]

The effect of crows and ravens on our culture is not one-sided. People also affect the culture of crows and ravens. This impressed one inquisitive and driven graduate student from the University of Wisconsin in the late 1970s. Hardened from a tour of duty in Vietnam, Rick Knight drove the back roads of rural eastern Washington and thought about ravens and people. He noticed that people harassed nesting Common Ravens more in farmland than in rangeland. In response, ravens were much more skittish around farmland nests than they were at rangeland nests. As Knight approached raven nests in rangeland, where people rarely destroyed them, parent birds remained on their nests until he got within a hundred yards of the nest site. As Knight climbed to the nests, the less fearful rangeland ravens approached to within a few yards of him, called frequently, and dove at him at a rate of nearly eight times per minute. In contrast to these bold rangeland ravens, farmland ravens, whose nests were often destroyed by local residents, left their nests when Knight got within five hundred yards and kept their distance, even when he climbed to the nest. Farmland ravens called half as frequently as rangeland ravens and rarely dove at the intrepid researcher.

Back in Wisconsin, Rick Knight noticed a similar difference between rural and urban American Crows. Rural crows have been persecuted by gun-toting farmers and sportsmen for two centuries, but urban crows have not. It is illegal to discharge a gun in Madison, Wisconsin. As a result, rural crows, accustomed to being shot at, call loudly as people near their nest trees but stay out of shooting range, even if someone climbs to their nest. By contrast, Madison, crows are quiet even if a person stands at the base of their nest tree. Should that person climb this tree, however, a crow riot will ensue, complete with aggressive swoops that will occasionally give the intruder a good wallop.[8]

Different reactions of crows and ravens to people clearly show us that the behavior of these corvids has been molded by their interactions with humans. We suspect that these different reactions also illustrate deep cultural differences that are evolving in crows and ravens as they encounter cultural differences in people. Persecution by people favors shyness and avoidance by crows and ravens. If shyness develops in some birds by watching other shy birds rather than experiencing at firsthand the wrath of people, then shyness can be said to have a cultural component. In the same way, if urban crows and rangeland ravens remain aggressive toward people because they see their parents and neighbors successfully chase unsuspecting and pas-

Nest defense at rural (left) and urban (right) crow nests

sive people from the vicinity of their nests, then aggressive nest defense is cultural.

Some background will help define what we mean by culture, why we are careful in applying it to corvids, and why we think human and corvid cultures are evolving together. One definition of *culture* is knowledge and tradition shared by a group that is transferred among individuals by some form of social learning. We have all experienced firsthand the variety of ways social learning accomplishes this feat. When your parents showed you how to behave at the dinner table, or if you copied your friends' tastes in clothing or hairstyle, or advertisements convinced you to buy a particular car, you were learning and adopting the culture of your social group. We diagram some of the ways social learning facilitates the spread of cultural traditions in box 1.

Many anthropologists and psychologists are reluctant to conclude that animals other than apes and humans have culture. Although we are comfortable identifying socially acquired behaviors as basic cultural elements, some scientists claim that culture consists of the set of rules that guide individuals, not the actual behaviors. These rules and conventions must also be encoded symbolically. Others accept that specific behaviors and rules are cultural artifacts but say that documentation of social learning producing them, especially the processes of teaching, shaping, and imitation, is incomplete. They argue that some behavioral traditions are too poorly studied to show how and if they persist through generations. They offer alternative ecological reasons, independent of social learning, for why some populations behave differently than others. We agree that culture is difficult to document in wild animals. But recent insights into the abilities and frequent opportunities animals have to observe and learn from others have convinced us that culture can and will frequently develop. When observant, intelligent social animals face changing conditions they may remain more closely tuned to their environment by evolving their culture than by relying on natural selection to evolve their genes. Culture can help animals adjust to their environment within and between generations, but natural selection must work between generations. Natural selection can craft elegant solutions to environmental challenges when conditions change slowly. Likewise, trial-and-error learning is necessary to meet challenges when conditions change unpredictably and often within an individual's life. But when conditions change at an intermediate frequency that demands adjustment within some generations and consistent behavior between other generations, then social learning is an efficient way to increase survival and reproduction.

Humans have changed crows' environments for hundreds of thousands of years. And the pace of that change is rapidly accelerating. To keep up with these new, and often dangerous challenges, we think that crows must respond culturally to people. This does not preclude genetic responses or individual learning. Genetic responses are expected and often underlie and coevolve with cultural responses. Cultural changes may even drive subsequent genetic changes as appears to be occurring with the evolution of lactose tolerance in people. Dairying is frequent in northern latitudes, perhaps because people with limited exposure to sun need to supplement their natural production of vitamin D. The human body normally produces vitamin D when it is exposed to sunshine. Dairy products also provide vitamin D, but many people cannot digest the lactose in dairy products. The genes that allow lactose to be digested have increased in northern people, in part because their dairying culture favored members of society with the genes encoding lactose tolerance.[9]

Social learning does not need to be a complex process. Nature provides ample opportunities for animals to observe each other's conspicuous activities. This "public information" is routinely used by animals to learn about new foods, new threats, new habitats, and even the quality of a potential mate. Observation and imitation of others is a simple form of learning that allows culture to develop, even if imitation is not perfect. Rudimentary culture may be refined substantially simply by parents guiding the clumsy or inaccurate imitations of their offspring. Some sort of social learning is required for culture to develop, however; simplistic trial-and-error learning is not sufficient. Trial and error teaches an individual how to behave, but it does not pass this information along to others.

Even though Rick Knight and his colleagues in Wisconsin have not clearly shown that the different ways rural and urban crows defend their nests came entirely from social learning, it must be part of the explanation. Natural selection ensues as aggressive crows are shot and cautious crows are spared. The population becomes more wary. But some crows cooperate and remain with their parents for years. This suggests that many naive crows probably learn how to treat potential nest predators by watching and mimicking their parents and neighbors (see box 1). Such social learning would be favored because it is clearly more adaptive than repeated trial and error "learning" about the range of a shotgun. Despite the plausibility of social learning in this case, it is difficult to demonstrate that inexperienced birds learn from their social partners (cultural transmission) rather than from direct experience (trial-and-error learning) without detailed observations of uniquely marked crows.

The ability to transmit culture through social learning equips social species like humans with a dual inheritance system. We obtain many physical traits, actions, and behaviors from the arrangement of nucleic acids on our chromosomes—that is, through inheritance of genes from our parents. But we also inherit a vast array of cultural traits from our society through social learning. Some researchers call the cultural units we inherit *memes,* to show the parallel between inheritance of genetic and cultural information. At any point in time, human culture is composed of memes that reflect genetic, individually learned, and socially transmitted information. Realizing that culture can be inherited allows us to conceptualize the process of cultural evolution. *Cultural evolution* is the change through time in socially transmitted behaviors that results because the behavior either affects the practitioner's survival or reproduction or it is chosen or imposed on others by conscious or unconscious decision-making. Archaeologists have documented both sorts of change in their investigations of the two-and-a-half-million-year record of human cultural evolution. The culture of horseback

Box 1 *Routes for the spread of cultural innovations
by social learning in people and corvids*

The arrows show how a behavior becomes a cultural trait as an individual (unfilled icon) learns about it from a knowledgeable individual (filled icon). In people (A), behaviors are transmitted culturally from parents to their offspring (i), among unrelated peers (ii), from teachers, leaders, or the media to members of society (iii), or from a collection of elders to younger members of a society (iv). (Adapted from Shennan 2002.) In corvids (B), cultural traits likely spread from parents and other members of extended families to young of the year (v; vocal dialects and the location and timing of local feeding opportunities may be passed on in this way), from the larger social group to unrelated crows (vi; communal roost locations, regional feeding opportunities, and migratory behavior may be reinforced this way), and from subsets of large social groups to the collective (vii; extended families may inform other nearby crows about hostile people or other new threats in this way).

Contact between crows and people could provide key interactions (dashed arrows) that fuel cultural coevolution of behaviors in both species (C). Ravens stealing drying salmon from a village elicit resistance from people (viii). As ravens lead their offspring and likely nearby unrelated birds to this new resource, resistance intensifies and spawns a culture of using scarecrows (ix). As ravens are increasingly harassed, they might pass a culture of wariness, stealth, and cooperation among their offspring and social partners (x). Finally, humans impressed with innovative ravens could pass on a culture of legends, myths, and taboos that create a mutually reinforcing living relationship between birds and people (x). The linear progress of interactions in this example is presented here for simplicity. Real cultural coevolution between people and crows may often be nonlinear, unpredictable, and chaotic.

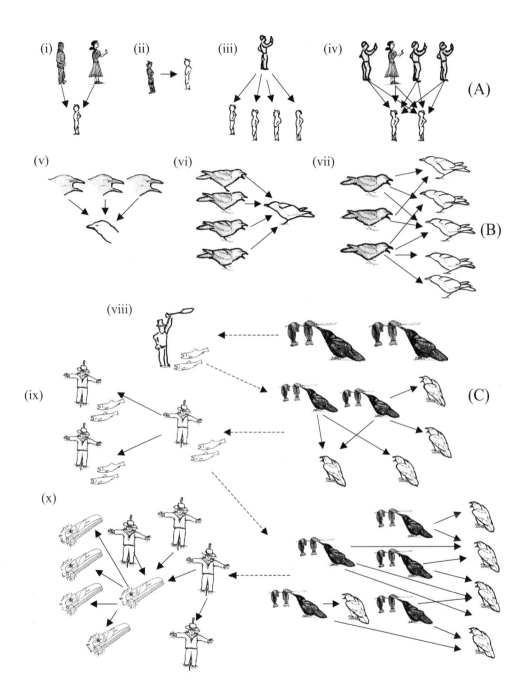

(i)　(ii)　(iii)　(iv)　(A)

(v)　(vi)　(vii)　(B)

(viii)　(C)

(ix)

(x)

riding, for example, evolved quickly in North America as people without horses procured less food or were killed more easily than those with horses. Similarly, the vocabulary used to define species, habitats, weather patterns, sea conditions, and other attributes important to fishers on Guara Island, Venezuela, evolved by conscious and unconscious decision-making.[10]

Because culture is transferred through teaching and observation, it can evolve and lead to predictable differences in behavior between animals that live in different groups. Much as people from different nations, towns, or tribes exhibit cultural differences, so, too, can animals from different flocks, troops, colonies, or herds differ culturally. Orcas (Killer Whales, *Orcinus orca*), for example, live either as residents or as vagrants and given their circumstances develop different hunting strategies and vocal behavior. Vagrant pods roam widely, silently hunting seals and birds, whereas residents live more sedentary lives and noisily hunt fish they have grown accustomed to taking. Perhaps in a similar way the culture of nest defense in crows and ravens has evolved to match human persecution. We would need to be sure that social learning is involved, but this supposition seems plausible.[11]

Humans are such cultural animals that we often take the uniqueness of culture and its transmission for granted. Although few would argue that other animals depend on culture to adapt to their world as much as people do, it would be incorrect to dismiss the presence and evolution of culture in other animals. Culture is a basic biological attribute of long-lived, social animals with well-developed memory abilities such as whales, dolphins, geese, and apes. Cultural evolution is especially likely to guide the practice of learned behaviors where mistaken actions are costly and environments are often changing. "Fishing" for termites by stick-wielding chimpanzees, complex song dialects of birds and whales, and migratory routes of Snow Geese (*Chen caerulescens*) are well-documented examples of culture

Vagrant Killer Whale grabbing
unsuspecting seal

in animal societies. These
skills are not innate and
do not merely spring into
being. On the contrary, developing these skills requires learn-
ing from observing others. The details of cultural development are particularly well known from studies of primates. In an unusually telling case,
island-dwelling Japanese Macaques (*Macaca fuscata*) were individually recognized and watched for many years. During the course of their studies,
researchers habituated the macaques by throwing food treats onto a sandy
beach. Imo, a two-year-old female, devised several innovations that eventu-

ally became cultural norms on her island. First she washed the sand off her sweet potatoes before eating them. Then she increased the efficiency of eating wheat by throwing handfuls of wheat and sand into the water and skimming the wheat off the water after the sand sank. Imitation of Imo's actions, particularly by other young monkeys, allowed her innovations to spread by social learning.[12]

We suggest that crows and ravens have culture and that cultural evolution is an important reason why they live so successfully with humans. Cultural evolution has likely played a role in the differences between European and North American corvids. Europeans have persecuted Hooded Crows, Carrion Crows (*Corvus corone*), and Common Ravens for centuries. As a result, ravens are rare and extremely shy in Europe, and crows are found in only a few major cities like Berlin, Warsaw, Oslo, Moscow, and Amsterdam. Crows have only recently invaded these cities, apparently proceeding under the cloak of winter's cold. Large flocks of crows join Rooks and Western Jackdaws to roost during winter nights in cities where the air is warmer and the people are more often inside and perhaps more concerned with the cold than with the annoyance large flocks of crows can bring. Eventually a few remain as breeders, where they enjoy the protection of new laws against shooting in cities and the lack of attention from most city dwellers who no longer see crows as a source of food. City crows eventually become more used to the new and less threatening urban people, and they join pigeons and sparrows in seeking handouts and scavenging from streets and gardens. But for every tame European city crow there are hundreds of tamer jackdaws. Jackdaws have been steadfast city residents with little fear of people. Their smaller size, less grating calls, mostly vegetarian diet, and infrequent predation on other birds' eggs or young has probably incited less human harassment. Jackdaws, crows, and Rooks are abundant outside European cities, but again their history with people is reflected in their present culture. Jackdaws have been persecuted the least. They are still easy to approach and hardly fear people. Rooks have been persecuted around their noisy nesting colonies and remain warier of people. Rural crows are still hunted and often persecuted because their diet includes sickly animals, birds' eggs, and the young of birds prized for hunting. Therefore, as in North America, rural European crows are extremely wary and quick to fly from people.

Our closeness to many crow species causes us to continually shape crow culture. Indeed, the speed with which we change the environment and the dangers we pose gives cultural evolution an edge over genetic evolution. Compared to the glacial pace of genetic selection, recombination, and mutation, cultural innovations spread quickly. They are also durable. Unlike individual learning, which is superior to social learning when conditions change often within an individual's life, cultural traditions allow individuals to gain knowledge from their contemporaries and ancestors. Just think of the variety of foods crows must have learned to recognize during the past century to flourish in a typical city: fruits and vegetables from around the world, cooked meats of all shapes, sizes, and colors, bread, pastry, noodles, ice cream, potato chips, sauces, and our personal favorite, the cheese puff. How did crows and ravens come to recognize fluorescent orange cheese puffs as food rather than bits of plastic? Derek Goodwin, a British authority on crows, similarly wondered about the ability of crows around the world to recognize bread as food. Because bread is fabricated and does not resemble natural foods, Goodwin suggested that crows learned to recognize and eat it, then passed this tradition through the ranks. They test new foods to see if they're edible, watch what fellow crows and people eat, and quickly incorporate an ever-changing variety of foods into their diet. At the same time, they avoid getting poisoned.[13]

English poet Ted Hughes summed up the essence of the crow's adaptability in his poem "Crow and the Birds" (1971):

> While the bullfinch plumped in the apple bud
> And the goldfinch bulbed in the sun
> And the wryneck crooned in the moon
> And the dipper peered from the dewball
> Crow spraddled head-down in the beach-garbage, guzzling a dropped
>   ice cream cone.

Another type of cultural innovation that may be important to crows is the use of tools. This is always a controversial concept among scientists, but New Caledonian Crows (*Corvus moneduloides*) have unquestionably evolved sophisticated tool manufacturing and use. Cultural evolution may be responsible for different traditions of tool use by these amazing crows. Although all New Caledonian Crows use a variety of tools to extract insects from decayed wood, and a propensity to use tools is partly innate, some populations specialize in the use of hooked tools while others use only straight skewers. American Crows have been observed to hunt by herding sparrows into the sides of buildings to stun them. In settled areas, crows drop thick-shelled nuts, clams, and tough-skinned squirrels on roadways, and then let the automobile traffic render their food. These novel techniques expand the wild crow's tool kit, which already includes twigs, splinters, and leaf stems used to extract spiders and insect larvae from crevices. The tool-using capacity of crows enables them to exploit novel feeding situations and profit even more from human society.[14]

A New Caledonian Crow fashions a tool from a leaf, transports it to the feeding site, and uses the tool to agitate a grub, which bites the tool and is extracted for consumption. The tool is kept and stored for later use.

Crows need to solve new problems quickly to keep pace with changes in the human environment. This may in turn put a premium on their ability to develop unique calls. Their calls vary regionally, like human dialects that can vary from valley to valley. Some of the regional variation in crow calls may arise culturally. When crows join a new flock they learn the flock's dialect by mimicking the calls of dominant flock members. This is cultural inheritance. Cultural inheritance of dialects might allow urban crows to quickly develop scolding calls for hawks that are different than those employed for cats; to perfect begging cries that motivate people to put more peanuts in the feeder; and to warn other crows of impending trouble. From our observations and those made decades earlier by Noble laureate Konrad Lorenz, it appears that humans who consistently harass local crows are eventually singled out and given special alarm responses by the birds. Whenever we walk across the University of Washington's campus, crows take our names in vain, warning all who know the language that the crow trappers are afoot. We don't know how they pick us out of the forty thousand folk scurrying like giant two-legged ants over well-worn trails. But single us out they do, and nearby crows flee while uttering a call that sounds to us like vocal disgust. In contrast, they calmly walk among our students and colleagues who have never captured, measured, banded, or otherwise humiliated them.[15]

Some aspects of crow and human culture appear to evolve together in a mutually reinforcing way, like the crow parading outside our window. When animals exert an evolutionary influence on each other in this coupled manner, we call it *coevolution*. Predators and prey coevolve as each selects for adaptations in the other. In this way prey may come to see more keenly or act more cautiously. A familiar, albeit more peaceful, example of coevolution is the relationship between hummingbirds and the fancy, tubular flowers they visit. The bill of the hummingbird has evolved its long length and decurved shape to reach deep into the flower's corolla and sip high-energy nectar. The flower has evolved hidden nectaries, deep within its corolla, to keep other organisms from stealing the sugary syrup and to ensure that any hummingbird that probes for nectar also gets doused with pollen. In this way, as hummingbirds go from flower to flower for food, they pollinate plants. By reserving the nectar for particular hummingbirds, the plant gains a reliable and highly mobile pollinator.

Hummingbirds and plants engage in a closely coupled coevolutionary dance. The shape of a flower evolves in response to a hummingbird's bill, and the bill's shape evolves in response to the shape of the flower. Coevolution can produce new flower or hummingbird species or even lead to differences in bill shape among male and female hummingbirds of the same species. On the Caribbean island of Saint Lucia, for example, male Purplethroated Carib hummingbirds (*Eulampis jugularis*) are smaller than females and feed on Heliconia plants with short corollas. The larger females have long, curved bills and feed on deep-flowered Heliconia. The deep-flowered Heliconia also contain more nectar than the Heliconia with shallow flowers, which helps the larger females meet their greater energy demands. Specialization on flowers that differ in shape has caused male and female hummingbirds to evolve different bill shapes. Reliance on small males or large females for pollination has caused flowers to evolve differences in nectar rewards.[16]

Usually we reserve the label "coevolution" for genetically based traits that evolve in response to interactions between organisms. We extend the concept of coevolution to situations where interactions among species lead to social learning and evolution of each species' culture. We call the coupled changes in two or more species' cultures that evolve in response to interactions between the species *cultural coevolution* (see box 1). Moose culture, for example, is modified by the culture of predators. Where bears or wolves prey upon moose, the moose react to wolf growls and howls and even the sounds of ravens, who associate with wolves. But where moose do not

Male (right) and female (left) Purple-throated Carib Hummingbirds on the Heliconia plant with which they have coevolved

live with their traditional predators like wolves and bears, they are also unresponsive to predator culture. Likewise, those vagrant Orcas mentioned above that hunt silently for seals may be exploiting the cultural coevolution that has occurred between resident seals and resident, fish-eating Orcas in Washington's Puget Sound. Resident Orcas have culturally derived dialects that pod members use to communicate noisily because salmon do not react to their clicks, whistles, and whines. The local seal culture has been shaped by ignoring the sight and sound of familiar fish-eating Orca pods that present no threat to them. Considerable time and energy is thus saved by Puget Sound Harbor Seals (*Phoca vitulina*) because, with sharks being rare here, they have been living a predator-free life for many generations. Occasionally, humans and the odd eagle take a young pup. But in 2002 a band of vagrant Orcas stealthily entered these waters and ate nearly half of the unsuspecting seals. These examples suggest that one species' culture can respond to another's culture. But is such cultural coevolution possible between people and animals?[17]

The Boran people of East Africa and a drab bird called the Greater Honeyguide (*Indicator indicator*) have closely intertwined their behaviors. The honeyguide is a specialized forager on the wax and larvae of wild honeybees. But despite its appetite, it cannot efficiently raid the hive of the honeybee. Hives are usually deep in tree crevices and vigorously defended by stinging swarms. The Boran people love honey, but they cannot efficiently locate the widely spaced and inconspicuous hives. So they work together, honeyguide and human. A Boran man, Hussein Isack, and researcher Uli Reyer measured how honeyguides adjust their flight pattern, vocal behavior, and perching behavior to actively, accurately, and consistently lead people through many miles of dry bushland to honey-laden hives. Birds reduce the search time of people for honey by roughly two-thirds. Because of this benefit, the Boran culture includes a specific and loud whistle, known as the "Fuulido," sounded by blowing into shells, nuts, or a closed fist, when a search for honey is to begin. The "Fuulido" doubles the encounter rate with honeyguides.[18]

Clearly the culture of following honeyguides by the Boran people has evolved by social learning—passing traditions along family lines and from elders to younger group members. But honeyguides and the Boran probably are not culturally coevolved. The behavior of honeyguides is not likely a result of cultural evolution. Honeyguides are not social species, and they are not raised by their kin. They are solitary nest parasites, like cuckoos, that are raised by adult birds of another species. Their abilities to guide people

Boran and Greater Honeyguide

and other animals to beehives most likely develops though individual trial-and-error learning or is mostly innate (purely genetic), because their solitary life and lack of association with parents precludes most sorts of social learning. Honeyguides teach us that intricate mutualisms among people and animals alone are not sufficient to infer cultural coevolution.

When humans interact with highly social species, cultural coevolution is increasingly likely. Bottlenose Dolphins (*Tursiops truncatus*) have fished with people at Laguna, off the coast of Brazil, since 1847. During this time dolphins have developed a culture of driving fish into fishers' nets, signaling to fishers, and feeding on stunned fish. This behavior is cultural because it is learned socially, persistent, and not found in all dolphin populations. Fishing with people has lasted for at least three dolphin generations. Young dolphins imitate or are taught by their mothers to herd fish, avoid nets, and signal fishers. The culture of fishing with people is not evident in all Bottlenose Dolphins, although it has evolved independently in several dolphin species that live in close association with fishing people. The culture of fishers includes an ability to interpret distinctive roll dives performed by dolphins. The intensity of these rolls, perhaps like the calling and flight styles of honeyguides, tell the fishers how many fish the dolphins are herding and where to cast their nets. Fishers rarely cast their nets until signaled to do so by dolphins. Coevolving cultures of herding and signaling by dolphins and interpreting and netting by humans allow both to fish cooperatively for mutual benefit.[19]

Few humans live in cultures closely coevolved with dolphins. If cultural coevolution is to regularly occur between people and animals, it most likely involves species we frequently interact with that are long-lived, intelligent, and social. In other words, crows! Relationships between crows and people are multifaceted (see the illustration on pages 8 and 9). Because people have such rich and variable cultures, we affect crows in many different, and often opposing, ways. These have changed through our history. As we have suggested, our first contact with crows, and especially with ravens, was probably extremely beneficial for these birds. Our hunting and gathering provided a reliable source of easily obtained food that they scavenged. We likely competed with ravens and crows for tens of thousands of years, during which time these clever corvids compelled us to improve our food storage abilities. Over a long period of interaction with crows and ravens, our ancestors came to admire the birds for their insightful skills and gave them a prominent place in our culture as sentient beings. They devised taboos to protect them, danced in their honor, wove them into stories for our children, and gave them handouts of food. Some peoples may have actually planted crops for crows. Early ethnographers from North America proposed this idea, indicating that the Makahs of the Pacific Northwest and other native people consistently associated crows with a honeysuckle that bore black berries (*Lonicera involucrata*). On first reading of this relationship, we assumed that it was the berries' black color that caused them to be known as "crow berries," but early conversations with native people indicated that the plants "were grown on purpose for crows." Perhaps they were cultivated by the Makahs or at least grown, metaphorically, for crows by their creator. Regardless, cultural coevolution was strong and mutually reinforcing for thousands of years. Ravens, and to a lesser extent crows, became increasingly reliant on people.[20]

This reliance changed as humans became agrarian. Now crows, Rooks, and Western Jackdaws interacted with people more frequently. Poor folk relied on sticks from crow nests to fuel their fires, but most people saw crows as clear competitors and developed a culture of harvest, harassment, and disdain toward them. In Europe, the scavenging of plague and battlefield corpses by crows and ravens amplified the growing local hatred and dread of the birds. As we have mentioned, this led to near extinction of ravens in Europe and dramatically altered the behavior of other species, such as Carrion and Hooded Crows.[21]

Humankind's nineteenth- and twentieth-century history as a mostly urban species has created a new haven for crows and, in recent decades, ravens. By creating a patchwork quilt landscape where small woodlots dot extensive fields of grass and crops, by excluding many native predators, and by piling our food scraps in reliable locations, we facilitate urban corvid life. In response, crows, Rooks, jackdaws, ravens, and other corvids, like magpies and some jays, are invading and increasing their numbers in our cities. Their cultures have changed to include new items in their diet, new substrates upon which to nest and roost, and new tolerance of a once-feared primate. This cultural revolution among crows is being met with awe, fear, and respect by people. It has been encoded in our popular culture. Outside cities we hunt crows for sport and food. Inside cities we use them as mascots and sources of inspiration. But their increasing numbers also cause us to consider lethal control. Crows have responded to our varied culture with a diversification of their own culture. Some are inextricably linked to that cornucopia of garbage, the dumpster, while others are established scavengers on less reliable offerings. The boldest birds beg at the windows of sympathetic urbanites. Still others inconspicuously raid hostile rural residents' feeders, gardens, and hunting grounds. These cultural adjustments may have produced the strong differences we detect today between rural and city crows. The greater length of time for cultural adjustment in Europe versus North America may be responsible for differences we detect today between cultures of crows of the New and Old World.[22]

Crows affect human culture more than many of the other wild species we reside with such as pigeons, starlings, squirrels, raccoons, or dormice because crows and people share fundamental biological and social properties. We're both gregarious, family grouped, long-lived, diurnal, vocally and visually astute, and reliant on memory and individual recognition. We are both generalists that use many different links in a food web. If crows can be thought of as specialists in any way, they are specialists on people. But does the commonness that specialization brings erode our feelings for crows? Are crows too common to command our respect? We think not. Some type of crow lives in virtually every city in the northern hemisphere, allowing crows access to hundreds of millions of people every day. Yes, crows turn off some people with their brash, boisterous attitudes, but by virtue of sheer numbers of encounters, we argue that the crows' subtle influence on us is more powerful than large wildlife icons like whales, wolves, cranes, and eagles. Rare species command the attention of modern policy makers as the public reacts to endangerment with interest and resources. But rarity, while having an impact, limits the global influence of these animals on humans to well below that of the crows.

Our cultural coevolution with crows may be unique because the utility of crows for people is limited. Our interactions with wildlife we eat or use in other ways certainly influences our culture, but this does not necessarily change the culture of these wild animals. Salmon, for example, have pervaded the culture of the people in the Pacific Northwest for thousands of years, but salmon still spend most of their life at sea before returning to their natal stream to reproduce. Wild salmon populations have declined, not flourished, at our hand. Some hunted populations of deer and elk have reduced their activity during the day or changed their movement patterns in other ways during the hunting season to be less conspicuous to people, but these species affect a relatively small portion of our culture.[23]

An extreme form of cultural coevolution exists between people and domestic species. Shepherd, sheep, and sheepdog have specific behaviors and signals passed on through tradition to increase humans' use of sheep.

Other notable examples of cultural coevolution between people and animals involve domestic animals. We have transformed the ecology, evolution, and culture of domestic animals away from their wild ancestors to meet our utilitarian needs, such as food, protection, and companionship. Many behavioral changes during domestication involve trial-and-error learning as we train our animals to respond appropriately. Other changes are surely cultural. A young sled dog learns to pull and take commands from the musher through training *and* by watching its teammates. Pure cultural evolution may be rare in domestic animals because our training and selective breeding produces the behaviors we want more quickly, more directly, and more efficiently than would occur if we relied solely on social learning to hone our herds, flocks, and teams.

Because domestic animals are no longer wild, we control what they learn and how they learn it. Their ecology, evolution, and culture respond only to people, not predators, competitors, or biophysical factors like climate. In this way, our cultural coevolution with crows again stands out as unique. Crows evolve in response to many aspects of the natural world, not exclusively to people. Crows learn from us what is useful to them. Useful behaviors are valuable and are therefore readily imitated and spread through social learning.

Coevolving species like crows and humans are always in a delicate, easily disrupted balance. Although our cultures may mutually reinforce interesting and novel behaviors, our ecological and evolutionary effects have more profound implications. We commonly shift the composition of native crow communities by changing the earth's landscape and even by moving species around the world. Some of our attempts fail, like those of the Russians who tried to establish Common Ravens on Alaska's Seal Islands to clean their fish camps. Others, like Rooks brought to New Zealand by homesick Europeans, succeed, and in so doing challenge the creative force of isolation that continental drift produced more than fifty million years ago. Actions like this often benefit some species but inhibit others. Our unintentional enhancement of some crow species and direct destruction of others is altering the balance of crow species on earth at an alarming rate. As we explain in chapter 2, our activities in the Americas are critically endangering unique island crow species and furthering the spread of successful mainland crows into regions not historically occupied. Such an influence may very well result in the extinction of distinct crow species and other birds, mammals, or insects linked to crows. In Wrocław, Poland, the escape of Hooded Crows from a zoo and subsequent explosive growth of an urban crow population, where none previously existed, has nearly wiped out the once-common native Wood Pigeon (*Columba palumbus*), as crows eat nearly every egg laid by pigeons. In short, we are slowing or stopping the creative engine of speciation while increasing the destructive forces of extinction. The result is a reduction in the diversity of native crow species specifically and of native birds generally. Our increasingly urban world will leave us with fewer kinds of crows and possibly more aggravation as settlement-savvy super crows expand their range and numbers.[24]

TWO

A Crow Is a Crow, or Is It?

While on a lecture tour in India in 1896, Mark Twain was besieged by flocks of House Crows (*Corvus splendens*) as he ate, smoked, and wrote. One can almost see the irascible author, his insights and thoughts stirred up by the horde of bold birds, as they pilfered his cigars and food and seemed constantly to evaluate and criticize him. Twain's description of the species as an accumulation of many incarnations perfectly suits all crows: "In the course of his evolutionary promotions, his sublime march toward ultimate perfection, he has been a gambler, a low comedian, a dissolute priest,

The corvid family; *clockwise from upper right,* Common Raven, American Crow, Black-throated Magpie-Jay, Steller's Jay, Clark's Nutcracker, Henderson's Ground-Jay, Black-billed Magpie, Piapiac, Collared Treepie, Western Jackdaw, Red-billed Chough, and Rook

a fussy woman, a blackguard, a scoffer, a
liar, a thief, a spy, an in-
former, a trading politi-
cian, a swindler, a professional hypocrite, a patriot
for cash, a reformer, a lec-
turer, a lawyer, a conspir-
ator, a rebel, a royalist,
a democrat, a practicer
and propagator of irrev-
erence, a meddler, an in-
truder, a busybody, an infidel,
and a wallower in sin for the mere love
of it. The strange result, the incredible result,
of this patient accumulation of all damnable
traits is, that he does not know what care is,
he does not know what sorrow is, he does
not know what remorse is, his life is one
long thundering ecstasy of happiness, and
he will go to his death untroubled, know-

Mark Twain contemplates his
nemesis, the Indian House Crow

ing that he will soon turn up again as an author or something, and be even
more intolerably capable and comfortable than ever he was before."[1]

One or another aspect of Twain's description resonates with each of us.
But few of us know that all crows are not the same. In 1926, American naturalist Frank Warne noted that "practically every inhabited land has a black-
feathered creature of the Crow genus, and though differing in size and habits
to some extent in various localities, climates, and amid varying surroundings, these birds are, as a rule, true to color—a glossy black—and whether

they are locally known as Crows, Ravens, Jackdaws, or Rooks, their instinctive sagacity, alertness, intelligence, and resourcefulness mark them as of the same species." Despite Warne's comments, the many diverse "crows" found worldwide are not of the same *species*. There are actually many species of crows, each with distinctive sizes, shapes, behaviors, and voices. These varied characteristics enable crows of one species to avoid breeding with crows of a different species. They also allow taxonomists, scientists who try objectively to classify living organisms, to recognize at present forty-six species of "crows" worldwide. These "crows" include birds commonly known as crows, ravens, jackdaws, and Rooks (see table 1 at the end of this chapter). Despite this splendid diversity, all crows share an important set of distinguishing characteristics that let them live throughout the world from the harshest desert and arctic climates to the most densely settled cities.[2]

The ancient Romans selected the raven as the head of all birds of omen, which they called the *oscines* (*os camon*). Contemporary scientists still classify crows and ravens (and most advanced singing birds) as oscines. But this is just the start. All crows and ravens are songbirds (order Passeriformes) belonging to the family Corvidae (crows and ravens plus jays, nutcrackers, and magpies) and the genus *Corvus*. Technically, *Corvus* includes crows, ravens, jackdaws, and Rooks.[3]

Most corvids have similar body shapes, featuring strong, stout legs with well-clawed, grasping feet, a robust midsection powering short, rounded wings, and large heads sporting distinctive bills. Individuals vary in size within and among species, but males average larger than females, and within mated pairs males are nearly always larger than and socially dominant to their female partners. Their stout, all-purpose bills are often likened to Swiss Army knives because they can cut, tear, crush, gape, probe, rip, and open just about anything. Bristles from the face, called rictal bristles, usually lay flat over the basal portion of the bill, covering the nostrils. The larger corvids walk rather than hop on strong, black legs and toes. This walk is often interpreted as a strut, with "a lordly and somewhat military air."[4]

All corvids have disproportionately large heads accommodating large brains that endow these birds with exceptional memory and intelligence. The brains of ravens relative to their body size are often claimed to be larger than those of any other bird. But this is not true. The pioneering Swiss zoologist Adolphe Portmann measured the brains of hundreds of birds in the 1940s. He concluded that, on average, corvids have larger brains relative to their body size than any other group of birds. But even though ravens topped Portmann's list of brainy corvids, the Green-winged Macaw (*Ara chloroptera*) and the Blue and Gold Macaw (*Ara ararauna*) topped the list of all birds. These large parrots' brains were two-thirds again as large, relative to their body size, as were the brains of ravens. Portmann should have measured American Crows, which among corvids have substantially larger brains than expected for their modest body size. Yes, dolphins and people have considerably larger brains relative to their body size than any bird, even the smartest corvid or parrot. But the relative brain size of crows and ravens is more similar to that of mammals, including most primates, than to other birds. Mentally, crows and ravens are more like flying monkeys than they are like other birds. This means that they are able to learn, remember, and use insight to solve natural and human challenges. Clark's Nutcrack-

A multipurpose beak is a defining feature of a crow. Here crows delicately carry eggs, tear open a bag, make a tool, place a nut before a car, ant, scrape and pinch a squashed bug from the pavement, crush a berry, flake bark, allobill, gape, and probe, pull, and hammer at a Moon Snail.

ers (*Nucifraga columbiana*) and Pinyon Jays (*Gymnorhinus cyanocephalus*), for example, store thousands of pine seeds each fall and, with an accuracy rate exceeding 90 percent, remember the locations so they can accurately retrieve the seeds from these nondescript, subterranean caches through the winter and spring, thus ensuring a source of food for the adults and young, which hatch early in the year. Pinyon Jays even learn from the experience of having their nests preyed on and choose less vulnerable nest locations. Ravens learn to pull meat hanging on a string toward them; Common Ravens and Hooded Crows even use this skill to tug on lines to retrieve hooked fish left untended by ice fishers. Surely pulling on a string to obtain a distant reward requires insight: the crow or raven must visualize the end result of acquiring food and carry out a complex plan consisting of pulling string a short distance, standing on it, pulling it further, standing on it, pulling it again, and so on to receive the reward.[5]

Bird brains are not entirely like our brains. In all animals the forebrain is responsible for learning and memory. But in mammals like us, the enlarged parts of the forebrain are the hemispheres with the distinct convoluted surfaces known as the cerebrum. The frontal and temporal lobes of our cerebrum are most important to memory, planning, problem solving, attention, and imitation. In smart birds like corvids and parrots, the inner portion of the forebrain, especially the nidopallium, is pronounced. The nidopallium is densely packed with nerve cells, making it an efficient storage, processing, and coordinating center. We know that the forebrain is important to learning and memory because it is largest in birds that excel in learning tests. Damage to it reduces a bird's learning ability. Another portion of the forebrain, known as the hippocampus, may be most important in spatial memory. In many corvids, chickadees, and tits (European chickadees) that cache seeds for future use, the hippocampus is huge. It may even expand

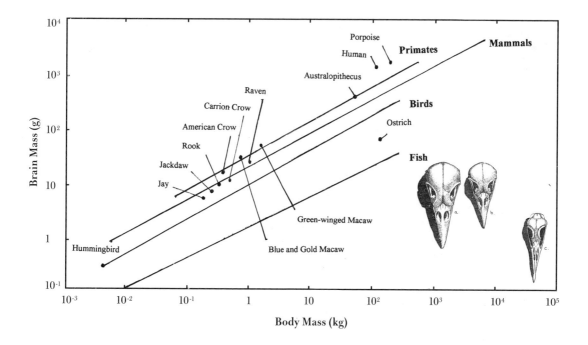

Brain mass increases with body size in animals. Corvid brains, relative to corvid body size, are more in line with expected mammal, even primate, brain sizes than with bird brains. But the braininess of large parrots, macaws, outshines even the raven. The relative braininess of corvids is obvious in the skulls of (a) Common Raven, (b) American Crow, and (c) Pelagic Cormorant. Despite similarity in body size to ravens, cormorants have small braincases. Skulls are drawn approximately ¼ life-size.

during the autumn when caching is common and shrink during the spring when stored foods are used less. Large forebrains do more than aid with spatial memory. They also allow corvids to discriminate finely among similar objects, live in rapidly changing environments, and enhance their social lives.[6]

Those who have raised or lived in close association with corvids often tell of habits learned by their birds to garner rewards or praise. McCaw, a raven raised by Tony Angell, was rewarded for returning from a neighbor's yard at the call of a whistle. On one occasion the bird brought a clothespin it had retrieved from the neighbor's clothesline. The next day, on being called by Tony, it sailed in with an even bigger prize—ladies' underwear. Some quick raven retraining was required, along with some awkward explaining to the neighbor. Bernd Heinrich, an astute student of raven mischief, was amazed by the story of a wild raven that carefully untwisted ten wires to open a squirrel-proof suet feeder in a Maine backyard.[7]

Anyone who has tried to catch a corvid knows that they quickly learn how to steal your bait without getting caught. Spending days cooped up in trucks or blinds waiting to catch a crow or raven makes your legs ache and mind wander. This led us to record the time it took for twenty American Crows to approach piles of white bread placed next to hidden traps. The traps were well-concealed net guns we fired remotely to catch crows. Ten crows had been trapped, measured, and banded previously and ten were presumably naive. These naive birds could have seen us catch other crows, but at least we had not caught and banded them. All twenty crows lived in our neighborhood, but naive crows approached our trap on average twenty-five minutes after we deposited the bread. Only one naive bird looked at the bread and hightailed it away. Perhaps we had missed him on a previous trapping attempt, or perhaps he learned through social means to avoid it. In contrast, 30 percent of experienced birds never even landed by the bread, and those that did took an average of forty-three agonizing minutes of sidestepping, staring, overflights, and aborted tries.

We think the crows' abilities to remember our traps are remarkable, especially given the type of trap we use. Our net gun is about two feet by two feet (less than a square meter) and easily camouflaged by vegetation. We fire it remotely using an electronic trigger and it goes off with the nerve-rattling bang of a .308 blank rifle cartridge. So a naive crow walks up to a nice pile of bread and *Blam!* a net flies out of the bushes and ensnares it. Then a couple of sprinting apes grab the crow, and poke, prod, measure, and band it before it is finally released. This humiliation lasts about twenty minutes. In spite, or perhaps *because,* of all that commotion, noise, and confusion, our crows remember the details of the food and trap well enough to avoid a similar situation a month and even a year later.

Memory of discrete events and fine details is important to a corvid's social life. Imagine how much time you would waste if you had to get reacquainted with your neighbors, your colleagues, your spouse and kids, or even yourself each time you met. This would be a large burden for any social animal, which is why most of them have fine recognition abilities. Ants recognize their colony mates by smell, many birds recognize their neighbors by song, and corvids, dolphins, and primates use their learning and memory abilities to discriminate familiar from strange fellows. Magpies (*Pica pica*), for example, can recognize themselves. Unlike the robin that spends all day defending its territory against its reflection in your window or the dog that barks at a mirror, a Magpie ignores its reflection unless it has an abnormality. Shine a red dot onto the magpie's breast, and it will carefully look in the mirror and then preen its own breast feathers in search of the apparent wound. A black dot shone on the black breast feathers does not produce such alarm. Self-awareness occurs only in the most cerebral species. We suspect it is just one of many demonstrations of the ability of corvids to recognize and discriminate social partners. Recognizing social companions allows crows and ravens to save time and energy by hiding food only from potential thieves; determining their dominance status relative to others without fighting; reciprocating only with their kin, mates, or past reciprocators; and gauging the extent of their mobbing responses to those in need.[8]

Crows and ravens are able to do more than just remember important events, locations, other birds, and people. They often appear insightful. For example, we often see crows and ravens take bread or other dry items to a nearby gutter or birdbath for a good soaking before they eat it. Ravens painstakingly cut large, but transportable pieces of frozen fat from a larger piece rather than simply chipping off small flakes. They also collect and stack up scattered crackers neatly in snow so that the entire pile can be transported rather than just one or two crackers. A favorite story of insight comes from U.S. Army Major Charles Bendire, one of America's first ornithologists, who described a pet crow named Jim that was teased with a knife. The knife was often brandished just out of Jim's grasp. Insightfully, Jim bit the teaser's hand, not the knife, causing the knife to be dropped. Jim was never taunted again by that blade, for the bird picked it up and flew off with the knife to a hiding place Bendire never discovered.[9]

The flexible behavior of corvids and their ability to use causal reasoning, insight, and imagination begs a comparison with apes. Despite differences in brain anatomy, corvids and apes have similar mental abilities. Like chimpanzees, who can use boxes and sticks to get a banana hanging out of reach, string-pulling Common Ravens show an ability to imagine an outcome. Tool-using New Caledonian Crows appear to understand causal connections, just as wild chimps do when they fish for termites by probing sticks into termite mounds. Caching corvids are deceitful and flexible. If a caching Common Raven is being watched, it will try to fake out the would-be cache robber with false probes into the ground or simply sneak out of sight. Apes are also flexible, often forming alliances or deceiving to achieve a desired outcome. Flexibility implies that corvids and apes understand general rules of association, like "cache to assure future food stores," not simple ones like "cache surplus food" or "cache when not hungry." It is not surprising that corvids and apes have full cognitive tool kits. Both live in complex and often variable physical and social environments where natural selection favors individuals that can learn, remember, and adjust.[10]

Members of the genus *Corvus* are usually wholly or mostly glossy black with some gray or white. This black plumage serves them well. Black feathers are stronger than less pigmented ones, and dark coloration both makes it easier for them to blend inconspicuously into the shadows to increase stealth or reduce predation and permits them to advertise themselves conspicuously against contrasting backdrops when they wish to emphasize social signals. Black birds absorb more solar radiation on sunny days than do light-colored birds, which allows them to conserve precious body heat in cold environments. This certainly helps on sunny but frigid arctic days and cold desert mornings. In the desert they also have the sense to forage in the open mostly early or late in the day when temperatures are tolerable and overheating is less likely. This has allowed the Hooded Crow of Europe to invade Cairo, Egypt, and explains how the Common Raven can live comfortably in the forests, grasslands, tundra, and deserts of both the Old and New World—fully half of Earth's landed surface.[11]

It thus seems that natural selection has yielded mostly black crows. The irony of this, however, has not escaped various peoples. Vietnamese folklore tells us that Peacock is responsible for Raven's black color. These two friends decided to paint each other one day. Raven went first and produced such a stunning masterpiece that Peacock was unwilling to share his beauty, even with his good friend, so he painted Raven black. According to Greek legend, Apollo turned the crow black after it told him about the marriage of his mistress.[12]

Regardless of the reason for blackness, it is not always complete. Partial

Box 2 *Vital statistics of our favorite crows*

Longevity: Banded (ringed) American, Fish, Hawaiian, and Northwestern crows live fourteen to twenty-four years in the wild. Common Ravens have lived thirteen years in the wild and forty to eighty years in captivity.

Breeding: Common Ravens are among the earliest breeding songbirds, often sitting on eggs in February amid deep snow. They lay three to seven eggs and incubate them for twenty to twenty-five days. Nestlings fledge at five to seven weeks' old. Crows lay three to nine eggs, incubate them for sixteen to eighteen days, and fledge young after four to six weeks in the nest.

Voice: Crows and ravens regularly give more than thirty distinct types of calls, but variation is great and true repertoires are much larger.

Adaptability: Common ravens live in the Arctic, where temperatures fall below -58°F (-50°C), and in the desert, where they exceed 113°F (45°C). Crows and ravens eat virtually anything, plant or animal. More than a thousand dietary items have been recorded.

Roosts: Raven roosts vary in size from fifty to two thousand birds each night. American Crows roost in groups of up to two million.

or complete albinism is regularly observed. We observed a wholly white raven for several years in southwestern Idaho. It so impressed our colleague Mark Pavelka that he dubbed it *Corvus clorox*. We routinely see American Crows with white feathers and even conspicuous, magpie-like white wing patches. Wholly white or uniformly grayish crows and ravens are rare because these sorts of "albinos" result from genetic flaws that render the birds

Flight Speed: Crows and ravens regularly fly at 30–60 miles per hour (40–100 kilometers per hour) and may attain speeds in excess of 70 miles per hour (more than 110 kilometers per hour) while diving at predators.

Home Range Size: Breeding American Crows use areas of 5 acres (2 hectares) to more than 500 acres (20 square kilometers). Breeding Common Ravens use from 245 to nearly 10,000 acres (1 to 40 square kilometers).

Migration: American Crows migrate up to 1,740 miles (2,800 kilometers), one-way. Ravens do not migrate, but vagrant Common Ravens have been known to travel more than 185 miles (300 kilometers) in a year.

Eyesight: Most crows and ravens likely have eyesight that is less binocular and less acute than people, but they detect objects with finer detail and more quickly than do people. Their color vision is well developed, and corvids may be able to see polarized and ultraviolet light.

More details can be found in the life history reviews by Pearson (1972), Boarman and Heinrich (1999), McGowan (2001b), Banko et al. (2002), and Verbeek and Caffrey (2002).

unable to produce pigment. This genetic variation is quickly removed from populations in nature by predators like hawks that routinely select odd prey. Partial albinos like the crows and ravens we often see with a few white feathers are more common because these abnormalities often result from injuries rather than genetics.[13]

The subdued plumage of corvids may select for another characteristic:

expression in voice, posture, and eye movement. Lacking the flashy color signals used by many other birds, corvids flash the whites of their eyes to signal aggression. They actually do this by blinking their protective, second eyelids, called nictitating membranes. Corvids also specialize in a wide variety of calls for most communication. They have specific calls for predators, mates, family members, feeding opportunities, and various needs, such as when young beg for their parents' attention. Not unlike the human voice, each call can be varied in pitch, volume, rate, intensity, and duration to convey subtle differences in the caller's mood or motives. Calls also convey individual, group, and perhaps family identity. To increase their expressive ability, calls are accompanied by posturing and body movement along with pupil dilation. Corvids also use subtle changes in the arrangement of their contour feathers to express emotion. People can easily read the mood of calling corvids when they position their bodies, just as do people and pets, in ways that clearly reveal fear, aggression, and need. This unbirdlike ability to express emotion so directly probably helps draw crows and people together. Watching crows perform can fire the artistic and spiritual imagination.[14]

Mating for life is another corvid hallmark. Mates perch, fly, walk, call, and often preen together. This dedication, however, is no guarantee that the same male fertilizes all the eggs in a nest. On several occasions we have seen "extra-pair" copulations at nests of Common Ravens in Idaho. These indiscretions occurred the instant a territorial male left his mate during the egg-laying portion of the annual cycle. Amazingly, as soon as the male left, another male flew in to copulate with the female on the nest. In some cases the interloper was a neighboring breeder. We do not know if these copulations resulted in fertilization, but it is possible, so we are careful about categorizing corvid pair-bonds as "genetically" monogamous. Functionally, there is

A mated pair of crows allopreens

no doubt about the strength of this bond. Mates groom each other by probing and combing their feathers to remove mites, flies, and other irritants, a
practice known as allopreening. This social activity is reminiscent more of
apes than of birds. Pairs function as highly synchronized teams, building
large, stick-based nests, carefully lining them with fine rootlets or hair, and
filling them on average with four bluish green, brown-flecked eggs. In the
Pacific Northwest, where huge cedars abound, crows line their nests with
finely shredded cedar bark. The natural insecticide that gives the tree its
pungent odor greatly reduces the occurrence of pesky nest parasites. This
practice may have been noticed by native coastal people, who relied heavily
on the tree, even fashioning clothes from its bark. The striking and seemingly incorrect coloration of crows' eggs also inspired several Pueblo Indian
legends. The basic thread is that the Pueblo people had to choose between a
beautiful, turquoise egg and a drab white egg. One egg held a parrot, while
the other held a crow. If the parrot egg was selected, good fortune would
follow. Yet people always picked the turquoise egg and suffered the consequences as a crow was hatched.[15]

Young corvids develop quickly. Incubation varies little among the many
species and usually lasts three to four weeks. Occasionally up to ten eggs are
incubated, but most pairs go for quality rather than quantity. Parents forage incessantly to feed their growing nestlings, which greet the world naked and helpless, a condition called altricial. Newborns weigh less than an
ounce but quickly grow to nearly adult size before fledging from the nest at
three to four weeks of age. Fledging refers to the act of leaving the nest and is
generally the time when nestling crows become independent fliers. We call
young crows fledglings after they can fly but before they are independent of
parental care, a period that lasts several weeks to months in most crow species. The fact that newly fledged crows resemble adults in size and color of-

Development of a crow from naked helplessness to confident flier: (a) five hours; (b) three days; (c) seven days; (d) eleven days; (e) fifteen days; (f) twenty days; (g) twenty-nine days. Based on photography by Ernest Good.

ten causes people to remark that they have never seen a baby crow. Certainly they have; they just haven't noticed it. Closer inspection reveals that young crows and ravens are less glossy than adults and have a pale blue iris that turns dark brown only after several months of life. Detailed morphological studies reveal that throughout their first year of life crows have shorter tails and wings than older birds, browner and more worn plumage, and distinctly rounded rather than squared tail feather tips. Parents care for their young for up to several months after they fledge. Fledglings closely follow parents, rapidly fluttering their wings and begging with a continuous gravelly whine so distinctive that we often use it to locate them. Some crows associate with their parents for years, even helping them raise subsequent broods. These "helpers" may in fact be primarily "learners," apprenticing to the tasks of feeding, tending, and protecting nestlings.[16]

Because corvids hold many strong distinguishing characteristics in common, many casual observers have difficulty sorting among the various species of "crows." In fact, living as we do in an area with crows and ravens, we are often asked, "How do I tell a crow from a raven?" Early naturalists must have been challenged with this same question. Our early ornithologist friend Major Bendire, for example, noted in 1895 that crows walk "more jerky" and not nearly "as dignified as the raven." Crows and ravens differ in many other respects. Ravens are large, often weighing more than 2.5 pounds (just over a kilogram) with prominent beaks, diamond- or wedge-shaped tails, and broad wings spanning more than 4.5 feet (1.4 meters). Crows typically weigh less than a pound (500 grams), with shorter and narrower beaks, fan-shaped tails, and wings spanning less than a yard (about a meter). Ravens often soar in flight, but crows usually flap, and their rate of wing beats is more rapid than the larger raven. Crows commonly flick their wings while also rapidly opening and closing their tails, especially after perching. Ravens also do this "wing-tail flicking," but less commonly than crows. The basso profundo *krawk* of the raven is easily distinguished from the comparatively anemic *caw* of the crow.[17]

Ravens live in a tremendous variety of habitats. They are the only bird that inhabits deserts, forests, scrublands, grasslands, taiga, *and* arctic tundra. They occur from the depths of California's Death Valley to the frozen heights of the Himalayas and from the northern Arctic to the southern reaches of Central America. Crows, by contrast, are rarely far from humans

American Crows pursuing their larger cousin, the Common Raven. Notice differences in tail shape, bill size, and throat plumage between the crows and the raven.

and typically live along coasts and coastal rivers or where forests intermingle with natural or human-made grasslands. The House Crow of India and southern Asia is found *only* with people. This species is expanding outside its natural range by hitching rides on our boats and following human trails. Our expanding roads and highways and conversion of forests, grasslands, and scrublands into agricultural fields, recreation sites, or settlements have enabled crow species to spread from coasts and natural forest-grassland interfaces throughout much of the world.[18]

We discuss the vocal behavior of crows and ravens in detail in chapter 6, but here we note some general features useful in distinguishing among species. Ravens have perhaps the most complex vocabulary of any bird. They *scream, trill, knock, croak, cackle, warble, yell, kaw,* and make sounds like wood blocks, bells, and dripping water. Their calls are hoarse and resonant. At a higher pitch, crows can *scream, rattle, whine,* and *coo,* but nine times out of ten, they *caw* (and *caw* and *caw!*). Both species are capable mimics. Variety and unpredictability define crow, raven, and other corvid calls, so whenever you hear something inexplicable in the forest or field, odds are that a corvid is the source.

The most ubiquitous and recognized corvid in the New World is the American Crow, *Corvus brachyrhynchos,* as named by Alfred Brehm in 1882. He focused on the difference between crow and raven beaks in his naming, as *brachyrhynchos* comes from two Greek words, *brachys,* meaning short, and *rhynchos,* meaning bill. *Corvus* is Latin for "a crow." In contrast, the raven's voice is highlighted in its scientific name (*Corvus corax*), as *corax* is derived from the Greek *korax,* which means "a croaker."

Taxonomists currently recognize the American Crow and nine other North American crows (plus two ravens) as full species. We catalog the salient qualities of these species in the table at the end of this chapter and illustrate their geographical ranges here. We do not know exactly how or when the nine crows diverged evolutionarily, but it has been suggested that following invasion of North America by an Asian crow, evolution proceeded separately on the mainland and islands. Ancestral crows colonized the continental United States and Caribbean but then underwent separate courses of evolution. In the continental United States, an early form of *Corvus brachy-*

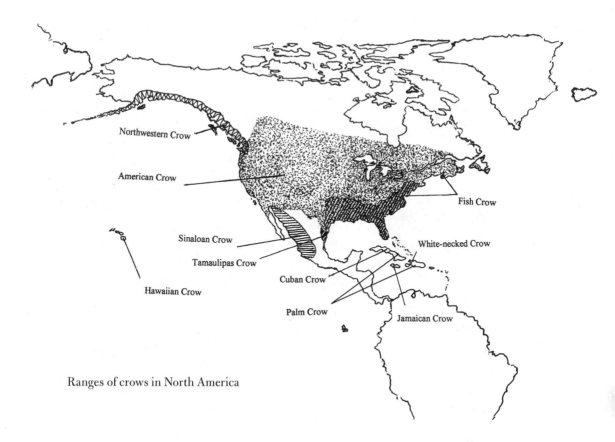

Ranges of crows in North America

rhynchos spun off the Northwestern (*C. caurinus*), Fish (*C. ossifragus*), Tamaulipas (*C. imparatus*), and Sinaloa (*C. sinaloae*) Crows. In the Caribbean, ancestral crows diverged on each major island into the White-necked (*C. leucognaphalus*), Cuban (*C. nasicus*), and Jamaican (*C. jamaicensis*) Crows. Finally, a more recently derived mainland crow, probably *C. brachyrhynchos*, reinvaded the Caribbean islands of Cuba and Hispaniola, diverging into the Palm Crow (*C. palmarum*), which was able to maintain its distinction even when sharing an island with the older Cuban and White-necked Crows. Genetic comparisons among European and North American crows tells us that the full creation of ten North American crow species occurred more than a million years ago, eons before humans arrived in these lands.[19]

Palm Crow

The taxonomy of North American crows has been controversial, mainly because of disagreement over how to, or whether we even should, distinguish among American, Northwestern, and Fish Crows. All three species occur in temperate North America. The Northwestern Crow of the Pacific Coast and the Fish Crow along the Atlantic now intermingle with American Crows, leading to claims that Northwestern and Fish Crows are merely subspecies of the American Crow. Looking for hybridization between suspected crow subspecies will be necessary to determine if in fact they are full species or subspecies. Hybrids are expected among subspecies but are usually rare among full species. Observing crows where species have the opportunity to interbreed may allow us to see evolution in action.

Crows nicely illustrate how isolation helps birds form individual species. Modern concepts of species are anchored to the notion that separate species are genetically distinct, a feature often maintained by an inability to interbreed successfully. New species of crows evolved only where they had the isolation necessary

White-necked Crow with toad

59

to reduce the likelihood of interbreeding with the widespread and abundant parental form. The unique and isolated beaches, tide pools, and especially islets and islands on the periphery of the widespread *C. brachyrhynchos*'s range were perfect microcosms for this. As crows spread throughout the Americas, those reaching islands adapted their behavior and morphology to unique foods of local forests. Glacial advances and retreats isolated Northwestern Crows in coastal refuges, where they adapted to reap the bounties of the sea. Mexico may have been settled by crow colonists from the Northwest who adapted to southern beaches, estuaries, and eventually human settlements. These evolved into Sinaloa Crows. Other emigrants from the Midwest or southern United States who took to the dry east brushlands of Mexico evolved into the Tamaulipas Crow. Isolation does not have to be geographic, a fact illustrated by Fish Crows. These birds may be able to maintain their distinctiveness while coexisting with American Crows not only by their unique vocalizations—they have a more nasal sound and say *uh-uh* in place of *caw-caw*—but because Fish Crows also lay eggs one to two months later than American Crows, effectively preventing breeding between the two species.

A few "founder" crows that colonize remote areas can grease the skids of speciation. By simple chance, a small sample of founder crows is unlikely to represent the full diversity of traits found in a widespread parental species. Perhaps the first colonists had slightly longer wings, duller plumage, or unique accents. These differences, if heritable either genetically or culturally, might persist as the population grew to fill its new home. This might explain why Jamaican Crows are dull, why White-necked Crows have white-based feathers, and why Fish Crows have long wings and nasal twangs. Island and coastal life also likely applied strong natural selection on crow settlers, favoring colonists with unique traits and skills. For example, eating clams might select for narrow, strong beaks or the insight to use tools, cars, or gravity to open tightly closed shells. Whatever the reason, once differences develop they are critical to speciation because they reduce the chances of interbreeding between otherwise similar animals.

Vocal differences are especially important for corvids, who communicate incessantly in this medium. Indeed, one can detect regional differences in crow calls, suggesting variations in dialects in the same species. Even slight differences in voice may be good enough to reduce breeding between diverging crow species. Mainland crows who arrived in Cuba after the Cuban Crows developed unique repertoires, for example, might not be able to understand the natives well enough to form pair-bonds and win mates. Not surprisingly, taxonomists have focused on crow vocal differences to

Fish Crow

distinguish species that now live side by side. When George Sutton, a pioneering ornithologist from the University of Michigan, first encountered Tamaulipas Crows on the east coast of Mexico, he called them handsome, but with an "odd, surprisingly feeble cry which resembled the syllables *garlic.*" Similar observations led to the classification of Northwestern Crows and Sinaloa Crows as unique species in the mid-1900s but also led others to conclude that Northwestern Crows were simply a well-marked subspecies of American Crows. This latter conclusion was based on an especially thorough study of crow morphology, behavior, and voice throughout the western coast of North America. Despite this work, current taxonomy bestows full species status on Northwestern Crows.[20]

Why do taxonomists differ in opinion? The simplest explanation lies deep within the working of natural selection. Natural selection does not create a species and then walk away. Rather, it continually shapes lineages—making them sometimes more and sometimes less distinct—through time. In a widespread, mobile species like a crow, portions of a population may diverge as they adapt to prevailing local conditions. This may lead particu-

Northwestern Crow at beach
edge with a scrounging relative

lar crows to form distinct populations and, eventually, whole new species. But as divergence occurs, individuals from the parental stock may swamp out any emerging distinction by repeatedly colonizing an area and breeding with the earlier, but not yet distinct colonists. Taxonomists cannot see this process through evolutionary time. They only have the briefest view of the process—a momentary glimpse at a process that has gone on for thousands or millions of years and continues still. Based on this instantaneous look they have to decide if the distinction is enough to warrant species status or simply subspecies status. In a widespread species inhabiting many distinct ecological settings in which populations are nonetheless well connected genetically through flight-powered dispersal, the divergence of subspecies into full species is often slow and incomplete. The taxonomist is left with a difficult puzzle that can be assembled more than one way depending on how the pieces are viewed.

Early naturalists who explored the Pacific Northwest saw firsthand the difficulty of separating similar species. Surgeons George Suckley and James Cooper, attached to the U.S. Army during the 1850–1870 survey work in the Pacific Northwest, were impressed by Northwestern Crows on the coasts. In their opinion, the crows "formed one of the marked ornithological features of the country." They noted that these crows were distinctly more gregarious than the American Crow they were familiar with. Unlike the American Crow, Northwestern Crows regularly nested on the ground, finding security in the cavities of the rocky coastal beaches. They were dependent on the sea and on Indian villages for food. In contrast, American Crows were not abundant and occurred only in the more open areas east of the Cascade Mountains and in settlements. Major Bendire reported in 1894 that American Crows were rare in the Puget Sound region in comparison to Northwestern Crows, which were common in the tidelands. He also noted that the distinction between Northwestern Crows and American Crows was much less than that between Fish Crows and American Crows. Today, the Northwestern Crow is purported to be smaller, to nest more colonially, to frequent beaches more exclusively, and to call with more nasal intonation than American Crows. David Johnston measured these attributes in crows from California to Alaska and could not find a cohesive group of Northwestern Crows that were distinct from their neighbors. Instead, he determined that each trait changed gradually from the south to the north. Alaskan crows sported the most "Northwestern" attributes, whereas Californian crows were the most "American," but Johnston could not say where the "Northwestern" variety took off and the "American" variety stopped. If there were two species in 1958, when Johnston did his study, distinct and abrupt changes in crow characters, not gradual ones, should have been detectable along the geographical divide between Northwestern and American Crows.[21]

These purported differences are also not striking to us. We catch crows together that vary from 10 to 15 ounces (300–450 grams), the range of both "species." We hear nasal and pure throaty calls from a single group of crows —sometimes even in rapid succession from the same individual crow. We have noticed separation of beach and inland crows, but these lines are continually blurred by human settlement and recreation in coastal areas. Humans link inland with coastal areas for crows by opening formerly isolating forests and providing food and nesting habitat to fuel crow population expansion. Therefore, our feeling is that although there may once have been a Northwestern Crow that was genetically distinct from the American Crow, that bird no longer exists. As noted Northwestern ornithologist Gordon Orians has put it: "The Northwestern Crow may have been swamped out of genetic uniqueness by interbreeding with its larger cousin, the American

Tamaulipas Crow

Crow." Interbreeding among crows is certainly plausible; it even occasionally occurs between crows and ravens. We suggest that this swamping was either facilitated or entirely enabled by humans. People have broken down the barriers between the two crows by modifying land that formerly isolated them and thus inviting both species into our settlements. Our sprawling, land-modifying housing settlements, highways, and waste-producing culture have become a thoroughfare and ever-expanding lunch counter for opportunistic species like crows. This helps explain why Johnston's detailed study of crows found only a continual gradation of size, behavior, and voice from southern to northern populations. The power of modern genetics will be required to determine definitively if swamping has occurred and whether our cultural activities have contributed to it. If our hypothesis is correct, it will demonstrate the powerful evolutionary force that humans have exerted on crows. We will have changed their course of evolution in a homogenizing way—forming a blended, generic, all-purpose crow where once there was distinction, uniqueness, and greater biodiversity.[22]

Even American Crows themselves show regional variation that humans may be reducing. Historically, taxonomists described four subspecies, or races, of American Crows: the eastern, western, southern, and Florida races. Eastern crows (*Corvus brachyrhynchos brachyrhynchos*) were widespread from southeastern and midwestern Canada to Texas and the middle portions of the eastern United States. This race migrates seasonally from parts of its northern range and has likely expanded west in the past several decades. The southern race (*C. b. paulus*) inhabits the southeastern United States and was originally described as smaller than the eastern crow and with a much slimmer bill. If differences between southern and eastern crows existed

Sinaloa Crow

or were emerging, they are not apparent today. The Florida race (*C. b. pascuus*) lives on the Florida peninsula and appears to have a smaller body and larger legs and feet than other races. Florida crows may also live in cooperative family groups more regularly than other races. Western crows (*C. b. hesperis*) reside throughout the western United States and southwestern Canada. They are smaller overall than the eastern race, may be more colonial in their breeding habits, and exhibit some migratory behavior from their interior, northern range. Western crows may have initially been a race of the Northwestern Crow, before the hypothesized swamping from eastern American Crows.[23]

Thus it is likely that one of the most profound and fundamental periods of interactions between people and crows is currently playing on the evolutionary stage. We continually sculpt the species we call crows by imposing strong selection on them to thrive in towns, cities, and agricultural settlements and enabling them to use the transportation corridors we construct between our settlements to expand their populations. Together, our activities favor contact between, and may allow for interbreeding among, formerly isolated lineages. Crows have responded to this selection in a number of interesting ways. Most profoundly, we suggest they are becoming smarter because learning, memory, and cultural evolution are so strongly favored by an increasingly complex urban lifestyle.

Picking a safe spot to sleep is one example of how crows use learning, memory, and cultural evolution to survive our challenges. Ohio crows demonstrated their ability to adjust roosting behavior to avoid human harassment more than fifty years ago. There, crows avoided hunters who staked out traditional roosts by congregating in scattered locations a few miles from the roost and waiting until near dark before flying en masse to the roost. Per-

haps such adjustment was just the first step toward a wholesale revamping of roost site selection by crows. Now, many rural crows roost in cities where hunting is not allowed. The ability to make such adjustments is continually favored by human unpredictability. In the not-too-distant past, crows that roosted near people would be eliminated. Now some are encouraged. The more unpredictable people are, the smarter crows must become.[24]

Crows may become more similar to one another as they evolve with our unintended but strong selection. Could we favor similarities among formerly distinct species to the point that interbreeding increases or a numerically dominant species overwhelms a more rare species? Along with the North-western Crow, American Crows might also reduce the Fish Crow's unique-ness if urbanization or global warming encourages eastern American Crows to breed earlier in the year, closer to when Fish Crows breed. This is already happening in many urban areas with a variety of corvids.

We also have extinguished some of the world's crow diversity. Many of the unique island crow species do not benefit as much from our pres-ence and changes to the landscape as does the American Crow. Populations of the Cuban Crow and Palm Crow are declining as agriculture and hu-man settlement remove the native forest habitat. Even more threatened, the White-necked Crow, Mariana Crow (*Corvus kubaryi*), and the Hawaiian Crow (*C. hawaiiensis*) have declined to the point where extinction is likely. Each of these crows is listed as endangered by the U.S. Fish and Wildlife Service. They are endangered because we have reduced their forest habitat, introduced exotic predators, and exposed them to new diseases to which they have little or no resistance. Indeed, the Hawaiian Crow, or 'Alala, is

Jamaican Crow in native bromeliads

Cuban Crow

one of the world's rarest birds. The last individual living in the wild died in 2002; fifty others, now in captivity, await repatriation. The Hawaiian Crow will likely follow the other two species of crows, known only from fossil records, that were extinguished by the first human inhabitants of Hawaii. As a result, Hawaii will lose a most eloquent native voice and its only endemic corvid.[25]

Species endangerment and extinction starts our story about how humans and crows form evolutionary convergences. People affect crow evolution more than vice versa. When crows tolerate us, they are assured of evolutionary success. But when a crow has needs that conflict with ours, it will likely be extinguished. As a result, the diversity of crow species that resulted from millions of years of creative natural selection is being reduced and the dominance of a few adaptable species is being enhanced.

Table 1. Crows and ravens of the world. We present all the known species in the genus *Corvus* but list those featured in our book first. This information is synthesized from personal observations, Goodwin (1986), and Madge and Burn (1994).

Common name	Scientific name	Distinguishing features
American Crow	*Corvus brachyrhynchos*	Voice, range, moderate to large size, close association with humans, bill smaller and tail more square than *C. corax*
Northwestern Crow	*Corvus caurinus*	Nasal voice, small size, musical flock calls, colonial nesting, range
Fish Crow	*Corvus ossifragus*	Nasal voice, late breeding season, some colonialism, small size, thin bill
Sinaloa Crow	*Corvus sinaloae*	"Ceow" call, range, long tail relative to wing, smaller and glossier than *C. brachyrhynchos* and *C. ossifragus*
Tamaulipas Crow	*Corvus imparatus*	Croaky voice, range, smaller and glossier than *C. brachyrhynchos* and *C. ossifragus*
White-necked Crow	*Corvus leucognaphalus*	Base of neck and contour feathers snow white, bare areas around eyes and bill
Cuban Crow	*Corvus nasicus*	Base of neck and contour feathers gray, bare areas around eyes and bill
Jamaican Crow	*Corvus jamaicensis*	Bare areas around bill and eyes, dull gray body plumage, small size
Palm Crow	*Corvus palmarum*	Base of neck feathers gray, short wings, distinctive voice
Hawaiian Crow ('Alala)	*Corvus hawaiiensis*	Large bill, distinctive calls, range
Common Raven (Northern Raven)	*Corvus corax*	Large size, distinctive voice, diamond-shaped tail, "Roman" nose

Habitat	Range	Current status
Farmland, forest-field interface, open woods, parklands, settlements	North America, southern Canada, northern Mexico	Expanding
Coasts, beaches, islands, rivers, settlements	Olympic Peninsula to southwest Alaska	Extinct? Likely hybridized with *C. brachyrhynchos*
Coasts, rivers, suburbs	Eastern U.S. coast, southeastern U.S. through Florida, west along major rivers to Oklahoma and Texas	Expanding
Coasts, rivers, settlements	Pacific Coast from Sonora to Colima	Stable?
Semi-arid desert scrub and brush-lands, settlements, agriculture	Gulf of Mexico coast from Nuevo Léon east to Rio Grande delta, south to Tampico, Tamaulipas	Stable?
Forest, pine woodlands	Haiti, Dominica, Puerto Rico	Endangered
Forest, wooded areas, treed settlements	Cuba, Isla de la Juventud, Grand Caicos Island	Possibly declining
Woodlands, parklands	Jamaica	Stable?
Wooded areas, treed settlements, agricultural areas	Cuba, Haiti, Dominica	Decreasing
Native moist Koa and Ohia forests	Island of Hawaii	Critically endangered
Forests, coasts, shrublands, tundra, desert, mountains, urbanizing lands, agriculture	Holarctic, south throughout middle Europe, Asia, and North America to Nicaragua	Increasing

Table 1 continued

Common name	Scientific name	Distinguishing features
Chihuahuan Raven	*Corvus cryptoleucus*	Flocking behavior, distinctive voice, large size, white-based neck feathers, range
Mariana Crow (Aga)	*Corvus kubaryi*	Range, small size
Western Jackdaw	*Corvus monedula*	Small size, flocking habits, gray nape, pale iris
Daurian Jackdaw	*Corvus dauuricus*	Dark iris, pied coloration
Rook	*Corvus frugilegus*	Bare, whitish face and base of bill, flocking habit, rookeries
House Crow (Indian House Crow)	*Corvus splendens*	Light to dark gray nape and breast, glossy black face
New Caledonian Crow	*Corvus moneduloides*	Tool use, range, solitary habits
Slender-billed Crow	*Corvus enca*	Smaller with longer, slimmer bill than *C. macrorhynchos*
Violaceous Crow	*Corvus (enca) violaceus*	Shorter bill and glossier plumage than *C. enca*
Large-billed Crow	*Corvus macrorhynchos*	Arched bill, large size
Jungle Crow	*Corvus (macrorhynchos) levaillantii*	Slightly smaller than *C. macrorhynchos*, square tail
Piping Crow (Celebes Pied Crow)	*Corvus typicus*	Range, white nape, breast and belly, screeching call
Banggai Crow	*Corvus unicolor*	All black, range
Flores Crow	*Corvus florensis*	Range, liquid and grating calls, all black
Long-billed Crow	*Corvus validus*	Range, large size, glossy plumage, long bill, white iris

Habitat	Range	Current status
Desert shrublands, low elevation woodlands	Southwestern U.S., northwestern Mexico	Stable?
Limestone forests, agricultural edges	Guam, Rota	Endangered
Towns and cities, farmland, open country, coasts	British Isles and western Europe, Scandinavia, northern Asia, northern Africa	Increasing
Towns and cities, open steppe, farmland	Eastern Asia and eastern Europe to eastern Japan, occasionally Scandinavia	Stable
Fields, open country, towns and cities	Europe, Asia, New Zealand	Increasing
Towns and cities	Indian subcontinent, Middle East, East Africa	Increasing
Woodland, open country, farmland	New Caledonia, Loyalty Islands	Common
Forest	Malaysia, Borneo, Indonesia	Locally common in forests
Forest	Philippines, Ceram, Moluccas	Likely subspecies of *C. enca*, uncommon to rare except on Palawan
Woodland, forest edge, rivers, villages, towns and cities	Eastern Asia, Himalayas, Philippines	Common, increasing
Forest edges, villages, towns, cities	India, Burma	Common, likely subspecies of *C. macrorhynchos*
Forest edges, open woodland, clearings	Sulawesi, Muna, Butung	Uncommon
Forest?	Banggai Island	Extinct?
Lowland forest	Flores Island	Threatened
Forest	Northern Moluccas	Poorly known, likely numerous

Table 1 continued

Common name	Scientific name	Distinguishing features
White-billed Crow (Solomon Islands Crow)	*Corvus woodfordi*	Large, arched, pale bill; range
Bougainville Crow (Solomon Islands Crow)	*Corvus meeki*	Large, arched, black bill; range
Brown-headed Crow	*Corvus fuscicapillus*	Brown head, large, arched bill; bills of males are black, females' yellow with black tip, juveniles' pale
Grey Crow (Bare-faced Crow)	*Corvus tristis*	Scruffy appearance, long tail, bare face, variety of coloration
Black Crow (Cape Rook)	*Corvus capensis*	Wholly black, thin bill, throat bulge like a raven's, pink eggs
Carrion Crow (Eurasian Crow)	*Corvus corone*	Wholly black
Hooded Crow	*Corvus (corone) cornix*	Gray nape, back, and belly contrast with black head and throat
Mesopotamian Crow (Iraq Pied Crow)	*Corvus (corone) capellanus*	Pied coloration
Collared Crow	*Corvus torquatus*	White nape and breast, black head, throat, belly, wings, and tail
Torresian Crow (Australian Crow)	*Corvus orru*	Small size, glossy black plumage, gray eye, bill longer than head
Little Crow	*Corvus bennetti*	Similar to *C. orru*, but smoother throat, bill shorter than head
Little Raven	*Corvus mellori*	Throat bulge not obvious, long wings
Australian Raven	*Corvus coronoides*	Prominent throat bulge, long wings
Forest Raven (Tasmanian Raven)	*Corvus tasmanicus*	Range, stocky build, heavy bill, long wings
Pied Crow	*Corvus albus*	Pied coloration, small bill, range

Habitat	Range	Current status
Hill forest	Southern Solomon Islands	Poorly known, likely widespread
Forest	Northern Solomon Islands	Poorly known, likely widespread
Lowland and foothill rain forests, mangroves	New Guinea	Rare, poorly known
Forest canopy and edges, villages	New Guinea and neighboring islands	Common
Farmland, open country, desert to alpine habitat	Eastern and southern Africa	Common, localized, increasing in Kenya
Settlements, farms, forest edges, clearings, coastal cliffs	Western Europe from British Isles to Germany, eastern Asia	Common
Same as *C. corone*	Northern and western Europe through Turkey, Syria, Iran, Iraq	Subspecies of *C. corone*, Common
Lowland agriculture, settlements	Southern Iraq to extreme southwest Iran	Poorly known, possible subspecies of *C. corone*
Lowland areas near water, farms, towns and villages	Eastern China, south into Vietnam	Locally common
Varied, but trees important, farms and settlements	Australia, New Guinea, and nearby islands	Abundant, increasing
Arid scrub, grassland and desert, farms and settlement	Australia	Abundant, declining locally
Arid, open country, towns and cities	Southeastern Australia	Abundant
Lush areas with trees, farms, towns and cities	Eastern and southern Australia	Abundant, increasing
Forest edges and clearings	Tasmania and adjacent south coast of Australia	Common, locally abundant
Open country, towns and cities, oases	Central African coasts to southern Africa	Common in settlements

Table 1 continued

Common name	Scientific name	Distinguishing features
Brown-necked Raven (Desert Raven)	*Corvus ruficolis*	Range, slightly smaller than *C. corax*
Dwarf Raven (Somali Raven)	*Corvus (ruficolis) edithae*	Similar to *C. ruficolis*, but diminutive in size, range
Fan-tailed Raven	*Corvus rhipidurus*	Batlike shape in flight, broad wings, short tail
White-necked Raven (Cape Raven)	*Corvus albicollis*	Large size, white nape patch, large bill
Thick-billed Raven	*Corvus crassirostris*	Largest extant corvid, range, massive bill

Habitat	Range	Current status
Desert, dry plains and foothills, oases farms, towns, and villages	South of *C. corax*, northern Africa, Arabia, southeast to central Asia	Common to abundant, especially in settlements
Same as *C. ruficolis*	Northeast Africa	Likely a race of *C. ruficolis*, locally common
Cliffs, canyons, plains, oases from below sea level to 13,000 ft. (4,000 m)	Northeast Africa, Middle East	Locally common, declining in Israel
Mountains, cliffs, gorges, hills, open mountain forest	Southern, central, and eastern Africa	Uncommon
Mountain ravines and cliffs, moorland, moist forest, farms and towns	Ethiopia	Common

THREE

Intertwined Ecologies
and Mutual Destinies

The Reverend Henry Ward Beecher once quipped, "If men had wings and bore black feathers, few of them would be clever enough to be crows." Today some might debate his claim, but few would argue that our earthly tenure of superior intelligence has been a long one. Corvids had at least an eight-million-year head start on people in the evolutionary race to intellectual superiority. This very likely meant that early people, who evolved from primate ancestors only about five to seven million years ago and did not experience a significant surge in brain growth until a mere two and a half million years ago, came face to face with savvy crows and ravens who could easily outwit them. Could this be why our ancestors so admired crows and ravens? Is this why corvids have so effectively exploited people over the millennia? Answering such questions requires a better understanding of the

Ravens in wait as some of the first people traverse the Bering Land Bridge to
North America

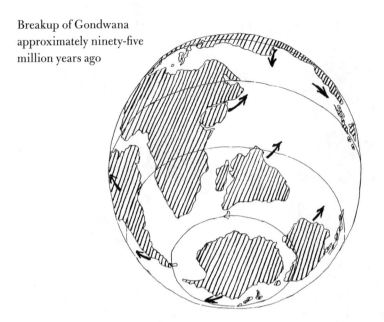

Breakup of Gondwana
approximately ninety-five
million years ago

history of ecological interactions between crows and people. It is a history
that reaches back far before the days of humans, to a time nearly one hundred million years ago, when Earth's original two large continents, Gondwana and Laurasia, were drifting apart.[1]

On the ancient southern continent of Gondwana, in the dark tropical
rainforests of what are now Australia and New Guinea, natural selection began to craft an amazing diversity of birds. Eventually more than five thousand species of songbirds would evolve and announce each new day with
their lovely and stirring chorus of song, but eighty to ninety million years
ago, things were just getting started. Australia and Antarctica, the remnants
of east Gondwana, were connected and drifting north into tropical climes.
Around sixty to seventy million years ago, Australia was moving north from
Antarctica into warmer isolation. This isolation on a large and diverse island allowed natural selection to create a bedazzling array of body shapes

and sizes, plumages, and behaviors among the early ancestors of songbirds. For the next thirty-five million years songbirds radiated into many of the forms we would recognize as prototypes of thrushes, wrens, tree creepers, jays, and crows. With island connections between Australia and Asia established twenty-five million years ago, these ancestors were free to colonize the north. They entered Asia and Europe and dispersed across land bridges to North America, Africa, and South America. Crows and ravens are relatively recent products of this storied evolutionary past. They diversified extensively in Australia, Asia, and Europe before a few representatives colonized North America. Their large body size, strong wings, and an ability to eat a wide variety of foods may have allowed them to travel more rapidly across new lands than most birds. Ancestors to modern corvids could have shown other songbirds the way from Gondwana to our modern world.[2]

As the first people crossed the fifty-mile-wide (eighty kilometers) Bering Land Bridge between Siberia and Alaska, the native crows and ravens were present and more than prepared to exploit whatever resources these travelers might provide. Whether these first Americans were hunters like the Clovis people or foragers on small game, fish, fruits, and nuts like the Athabaskans of Alaska's Nenana region would not have mattered to the native corvids. Crows and ravens would have pilfered the leftovers of either diet. Likewise, even if early Americans began to arrive thirty thousand years ago and came from several routes to western North America, crows and ravens would still have been waiting. They had already perfected the habit of cleaning the waste of hunting, gathering, and scavenging mammals. Now they had met the most wasteful of all, one that often killed more than it could eat, discarded sizeable proportions of food as unpalatable, and was capable of transforming the earth's surface in a manner that favored the basic needs of these birds. Of course we cannot be certain about this early history, but the genetic, fossil, and anthropological record does allow for some cautious speculation.[3]

We know that the early ancestors of modern North American crows and ravens likely crossed the Bering Land Bridge before humans. Although there could have been repeated invasions of the New World by songbirds from Asia across the Bering Land Bridge as early as fifteen million years ago, genetic studies suggest that corvids likely colonized later. Molecular geneticists have begun to unravel some of these mysteries by looking at the similarities and differences in the genes of modern crows and ravens. Rob Fleischer of the Smithsonian Institution compares the sequence of building blocks in the DNA of American Crows to that in other corvids' genes. Fleischer is a gentle, unassuming scientist who carries in his laptop data on several hundred sequences of the nucleic acids that compose the DNA backbone of a crow's mitochrondrial genes. With a click of his mouse, he can calculate the similarity between various crow's genes and thereby reconstruct their evolutionary relationships. He and his colleague Carl McIntosh have determined that American Crows are more closely related to European Carrion Crows (*Corvus corone*) than they are to five other species of crows, Rooks, Western Jackdaws, and ravens. American and Carrion Crows share about 95 percent of their genetic material, which suggests that they diverged from a common ancestor more than two million years ago.

Fleischer and McIntosh did not measure all the crow genes to make this inference. Rather, as is usually the case in this sort of work, they sampled a specific gene, Cytochrome *b,* that is found in the mitochondria within a crow's cells. Mitochondrial DNA is not strongly under the pull of natural selection and changes through time as random mutations accumulate. This "molecular clock" takes about one million years to accumulate a 1.6 percent difference in DNA in songbirds. Most avian geneticists feel that the rate falls between 1 and 3 percent. Two to three million years ago the Bering Land Bridge was covered with extensive coniferous forest. Bogs and deciduous woodlands were also present. This is typical of Carrion Crow habitat in northern Europe today, so it is plausible that ancestors to modern Carrion Crows and American Crows would have readily crossed the land bridge at that time. Land bridges also formed at least three times during the last five hundred thousand years, but then they were covered with grasslands that might have been a less appealing habitat to crows. Geneticists have yet to begin to investigate the relationships among the extant North American crows, but we suggest that the ancestral American Crow diverged into Northwestern Crows and Fish Crows after reaching North America from Asia.[4]

The fossil record is consistent with these early speculations and confirms the presence of crows in America well before humans. Steve Emslie, a paleontologist from the University of North Carolina, has crawled through many caves and sifted their sand floors in search of bird fossils. From his work, we know that crows were found throughout North America one to two million years ago. Emslie's persistence allowed him to discover that American Crows, Fish Crows, and at least one extinct crow were well south and east into North America early in the last Ice Age. That is, they were in Florida approximately two million years ago, during the Pleistocene. Crow fossils also were found at Porcupine Cave, Colorado, dated to the early and middle Pleistocene, some 1.6 to 0.67 million years ago. Emslie's discoveries suggest that crows frequented much of ice-free North America for millions of years. They likely exploited natural fruits, grains, insects, eggs, nestlings, and small vertebrates, scavenged the remains of predators' kills, and reaped the sea's bounty from its intertidal and coastal tables. An exception to their wide distribution may have been in the southwestern United States, where Pleistocene crows are unknown and colonization of desert areas has been slow even in modern times. Emslie's early fossils suggest that crows must have crossed the Bering Strait well before the start of the Pleistocene to disperse the 4,350 miles (7,000 kilometers) south to Florida one and a half million years ago.[5]

Ravens may have preceded crows into North America. Their larger size and stronger flight make them ideal candidates for early and frequent land bridge crossings. The presence of wolves and their ancestors, which also traveled across land bridges, may have aided the raven's movement by providing frequent scavenging opportunities. Ravens and American Crows have evolved separately for about four million years, probably because ravens diverged from a common European and North American crow ancestor in Europe. Genetic evidence suggests that ravens have been in North America for at least two million years because at that time the Common Raven began to diverge into two well-distinguished genetic forms. Kevin Omland, a former ski racer turned geneticist at the University of Maryland, ground up hundreds of raven feathers, extracted their DNA, and made a startling discovery. Not all "Common Ravens" are the same. The most common form, known as the Holarctic Clade, is found throughout the northern hemisphere in Europe, Greenland, Asia, and North America. But the less common California Clade is found only in North America, primarily along the southern Pacific Coast. Common Ravens must have been in southern North America two million years ago to participate in this divergence. At that time, southern ravens were probably isolated from their European and Asian counterparts when massive glacial ice sheets covered the northern world. Later, Holarctic birds once again crossed into North America, perhaps flanking wolves and people. Today the clades are merging and distinctions may blur. But about one million years ago the California Clade was sufficiently abundant to produce a second actual raven species, the Chihuahuan Raven (*Corvus cryptoleucus*). In so doing, ravens or crows, and often both, inhabited nearly every part of North America well before people arrived.[6]

When Siberian hunter-gatherers crossed the land bridge thousands of years ago, they may have been accompanied by Holarctic ravens and contacted crows along the Bering shore or in the scattered woodlands further south. Fossil ravens are found in the earliest known human camps in western Canada, documenting a close relationship between ravens and North American people for more than ten thousand years. Their dogs, domesticated from European wolves, would have been especially attractive to ravens, which were closely associated with wild wolves. Perhaps coastal crows along the north Pacific rim remained with these Siberian hunters for a few thousand years before people diverged into the many ecological zones of the Americas. Coastal peoples likely associated with Common Ravens, Northwestern and Fish Crows, and Clovis people certainly found themselves in the company of crows and ravens as they started hunting the abundant big game of North America.[7]

These early associations between crows and people may have reinforced or even enhanced differences among the species and subspecies of crows in North America. Native Americans differed in diet and in the form and persistence of their settlements. These qualities may have selected for some of the differences in morphology we see among Northwestern, Fish, and American Crows today. At least it's an intriguing thought, but it will have to await more genetic analysis and fossil dating.

P eople are a potent though inconsistent ecological force on corvids. When we humans were uncommon and isolated in the productive regions of North America, it is likely that our pantheistic view of nature kept us from harming crows or ravens despite their fondness for our drying fish. Indeed, the reverence for crows and particularly ravens reached an extraordinary level of cultural refinement among the early peoples of the Pacific Northwest. Such respect or at least tolerance likely increased crow and raven birthrates while lowering their death rates, thereby increasing their ability to compete with other species. Our dietary differences may also have helped our three North American crow species diverge, as suggested above. This ecological honeymoon between people and corvids would soon change, however. European explorers and colonists had been forming different attitudes toward crows and ravens and would therefore offer new ecological and evolutionary challenges to crows when they arrived in North America. As much as modern people have been inspired by crows to dream, sing, dance, and paint, this inspiration has waxed and waned conspicuously in Europe. The Paleolithic people who sought shelter fifteen to thirty thousand years ago in the caves near Lascaux, France, decorated the cave walls with images of corvids, and the beneficial aspects of crows, Rooks, and Western Jackdaws were appreciated through the late 1400s and early 1500s in England. There, corvids were legally protected from destruction because of the janitorial services they performed on city streets. A special decree in 1534 issued by King Henry VIII also protected corvids hunted for sport by falconers. English ravens were protected in the 1500s because of their valuable service as scavengers of putrid meat. Crows and ravens signaled longevity from atop sheaves of grain in family crests of this period. But inconsistency was evident. Henry VIII's protection did not extend to the valued corvids' eggs. Bounties were developed to help farmers rid their crops of bird pests. Even Henry's per-

The coat of arms of the Harry Holmes-Tarn family depicts a raven on a sheaf of wheat. The banner proclaiming "eternal life" respects the raven's longevity and positive association with agriculture.

sonal connection to corvids was severed when he had his second wife, Anne Boleyn (aka "midnight crow"), beheaded.[8]

Less than a century later, Britons' appreciation of corvids had turned to hatred. Crows and ravens' habit of feeding on corpses from plagues and battlefields gave them a reputation as harbingers of death. The devastating London fire of 1666 sealed many birds' fates: ravens and crows scavenged the London dead with such vigor that survivors of the fire were shocked into asking King Charles II to exterminate them. Bounties were placed on crow and raven heads. Nests were destroyed. British attitudes now resembled those of peoples farther east. Muslims were intolerant of crows during the Middle Ages and Renaissance. In Iraq, crows were considered one of the five scoundrels of the animal world. People were allowed to kill such villains, even while on pilgrimage to Mecca.[9]

Corvids in general were viewed as "vermin" in Europe throughout the 1700s, 1800s, and early 1900s. In western Europe, ravens were driven from cities and hunted with such passion in the countryside that a "cultural gap" in their geographic distribution developed. Though they survived in Scandinavia, northern England, Spain, and southern and eastern Europe, ravens were extirpated from most of Germany, Poland, the Netherlands, Czech Republic, Hungary, Slovakia, Austria, and southeastern England. Some persecution was reduced in England and Scotland in 1981, when ravens received protection under the Wildlife and Countryside Act. Common Ravens remain protected today. Their lesser brethren and direct ancestor to the American Crow, the Carrion Crow, did not receive protection and to this day is hated and persecuted by many gamekeepers, farmers, and shepherds in the United Kingdom. Its sibling species, the Hooded Crow, fares no better in Scandinavia and much of western Europe.[10]

These attitudes likely accompanied early European settlers of North America. English settlements on the Massachusetts shore certainly attracted American Crows and possibly Fish Crows. Increasing European colonization of North America probably sped the association crows had with people. Settlers may have viewed these crows as vermin and competitors for limited game and crops. They quickly established bounties to encourage removal of "blackbirds," which likely included crows. Pennsylvania settlers were required by law to shoot a dozen crows to claim land on the frontier in 1754. In this way the general symbiotic relationship between North American corvids and early peoples that had persisted for at least twelve thousand years was abruptly shaken by Europeans beginning in the 1500s. Over the next three hundred years this changing attitude toward crows and eventually ravens gradually established itself as the cultural traditions of native peoples were displaced by a Eurocentric culture. During the twentieth and twenty-first centuries, North American crows have lived with increasingly hostile humans, few of whom appreciated their services and many of whom actively attempted to destroy them.[11]

Fearful of disease, convinced of their negative effects on game and grain, and annoyed by their noise, Americans began to wage war on crows in the mid- to late twentieth century. Persecution of ravens was not conspicuous during this period because ravens did not live in close company with most of the population. In the 1930s and 1940s state fish and game agents used dynamite to bomb crow roosts in many midwestern states, killing hundreds of thousands of crows. Bombs killed 18,000 and 26,000 crows, respectively in a single night's blast in Binger and Dempsey, Oklahoma. In the winter of 1939–40, the Illinois Department of Conservation killed 328,000 crows at roosts. These attempts to eliminate crows may have lessened local crow "problems," but they had no effect on regional populations, which continue to thrive throughout the Midwest today.[12]

Killing the occasional pesky crow may be justified, but many crows are killed annually simply for the fun of it. By midcentury crow hunting was a big sport in much of rural America. Its popularity prompted Bert Popowski to write *The Varmint and Crow Hunter's Bible* in 1962. Popowski encouraged hunters to stalk crows because of the challenge they presented, noting that crow hunting required stealth, skill in calling and shooting, and perseverance. Crows made excellent targets at all times of the year to sharpen shooting skills, he wrote. And they could be shot when other targets were unavailable. Skilled crow hunters have killed hundreds of crows daily, thousands annually, and tens of thousands in a lifetime. By the time he wrote his book, Popowski's personal crow killing had clearly reached epic proportions; he claimed to have taken eighty to ninety thousand himself.[13]

Crow hunting is enigmatic. It differs from other hunting sports in three distinct ways. First, the goal is to kill as many crows as possible, not to experience the thrill of the chase with a specified bag limit. Second, crows that are killed are not eaten. Third, the quarry is hated, rather than revered or respected, by the hunter. That Popowski would title his book a "bible" is fairly ironic, for there is no reverence to be found in its pages. Hatred of crows rolls out of this book, and new levels of anthropomorphism are achieved as the author refers to young crows as "stupid," to all crows as "loud mouthed cowards," and to the only "good crow" as a "dead crow." He also constructs an unusual simile as he compares crows to evil characters or political enemies of the past century, likening crows to Adolf Hitler, Benito Mussolini, Mao Zedong, and Fidel Castro because they can whip a crowd of followers into a frenzied, murderous gang. Hatred of crows by hunters sheds some light into the power crows hold over their assailants. Perhaps crow hunters need to devalue their targets to avoid the guilt and shame that usually accompany senseless killing. Or are crow hunters just blaming crows for problems that go well beyond what the birds are actually responsible for?

Certainly, in our experience, crow hunters seem to think that being able to shoot crows is too good to be true. One cold November evening while tracking radio-tagged crows in Idaho we heard gunshots near the roost. Wounded and dying crows rained from the sky and dead ones covered the ground. The hunters were apologetic when we explained our research objectives to them. Even though their shooting was perfectly legal, they assumed they had done something wrong.

We have seen that humans responded to corvids over the ages in opposing ways. Originally, crows and ravens inspired us and serviced our sanitary needs. They fell from grace in Europe before Europeans colonized the Americas. Early Euro-Americans thus held crows in low esteem, an attitude shared by many modern Americans. Interestingly, the larger, more solitary, and less commonly encountered cousin of the crow, the raven, enjoys wide respect today. Perhaps crows have not gained our appreciation like ravens because they are simply too common. Commonness may engender lack of respect and distrust and fuel the feeling that wanton killing is all right. Regardless of the reason, our inability to decide in favor of or against crows has put them in an unusual legal position. On one hand, they are protected by federal law in the United States under the Migratory Bird Treaty Act. On the other hand, they are regularly controlled by federal agents and hunted simply for sport during seasons when other targets are unavailable. It seems to us that crows demand respect simply for being such tenacious survivors despite persistent persecution.

In the past hundred years North Americans have become increasingly urban. Cheap fuel, scuttled railways, aggressive marketing by the auto industry, and expansion of road systems created suburban sprawl and made it the defining feature of the American landscape. The land area of our cities has doubled or tripled in the past fifty years alone. By converting farmland, forestland, and wetlands into savannas of grass and exotic trees, we have made a perfect landscape for crows. Our neighborhoods surround and perforate forests once inhospitable for crows. Our desert cities expand the once rare or seasonal riparian areas that were the crows' only toehold in these hostile lands. Small hobby farms and luxury homes that provide food, cover, and tolerance for crows have replaced many of the extensive ranches in the intermontane West. In sum, we have transformed North America into an Eden for crows with a nearly perfect arrangement of nesting and feeding habitat. To sweeten the deal even more, we restrict shooting and most persecution of wildlife in cities.[14]

Crows have responded in a huge way. Crows are nesting and roosting in rising numbers in urban areas across the United States. These increases are closely associated with increasing human populations across the country, especially in western, coastal cities. Crows have expanded into new regions of the West, where only fifty years ago no crows were found. Now more than ever, American Crows are urbanites. They are rarely found far from humans and reach their greatest densities in our cities and suburbs, where there is ample food and shelter. This close association with humans gets them noticed and called a variety of names. In scientific language we call them human commensals or synanthropic birds. Crow fanatics refer to them as lofty messengers or incredible problem solvers. Many city dwellers see them as noisy trash birds. To lots of people, they are simply "those damn birds that wake me up each morning." But to the homeless people in at least one American city, they are a true inspiration. A group of proud Seattle people living on the streets call themselves the Tribe of Crow. Their pride is in contrast to many homeless folk and stems from their ability to put culture's waste to use, finding what they need discarded all around them and thriving by sharing abundant, untapped resources. And these characteristics are indeed what make crows such successful urban birds.[15]

American Crows are not unique as urbanites. Around the world, other corvid species are exploiting humans and increasing in abundance. Rus-

Distribution, current population trends, and recent expansion of American Crows in western North America. Southern British Columbia, Washington, Oregon, Idaho, California, and Nevada are outlined. Symbols indicate new sightings (range expansion) and population trends as indicated by annual Christmas Bird Counts (after Withey 2002).

○ decrease
● increase
◌ no change
▲ range expansion

200   0   200   400 Kilometers

sian scientists report jays, magpies, and Carrion Crows on the rise in Moscow. Italians have documented an increase of Western Jackdaws in Rome. Magpies are increasing throughout Europe and dominate cities as diverse as London, Oslo, Berlin, and Warsaw. Jungle Crows are a pestering nuisance in Tokyo, where the metropolitan government spends millions of yen (more than ten thousand dollars) annually funding citizen groups to assess the problem and professional crow exterminators to act immediately. In spite of such efforts, Jungle Crows prompted nearly thirteen hundred complaints in 2000 for noise, scattered garbage, and violent attacks on people.[16]

Seattle and the surrounding Puget Sound area of western Washington has been one of the fastest growing urban areas in America in recent decades. A human population of 1.5 million supports a large and growing crow population. Here crows do pretty much what they want throughout the city. They walk boldly down inner city streets, beg from autos at ferry landings, plunder garbage dumpsters, and wait eagerly outside every fast-food restaurant for their next meal. They are regularly uninvited guests and an insouciant presence at picnics. It has not always been like this. Rather, crow populations have skyrocketed in response the growing human population. Nearby wildlands in the Cascade Mountains and on the Olympic Peninsula allow visitors to Seattle to take a time-machine ride back to view primordial crow populations. Crows become less abundant as we move from urban to wildland settings, and their populations in small cities beyond the Puget Sound have not expanded as have those in Seattle. Virtually the only place you can find crows in wildland settings is along ocean coasts or near resorts, campgrounds, and clusters of vacation homes.[17]

To understand why crows prosper in urban settings, we have captured, tagged, and observed crows for seven years along a gradient of lessening human presence from Seattle to Olympic National Park. We quickly learned that city crows use less than one-tenth of the territory that wildland crows use. City pairs defend areas no larger than a few house lots to supply their needs, but wildland crows need several square miles to garner enough resources to live and breed. In addition, wildland crows regularly make extensive trips of 5–20 miles (10–30 kilometers) to areas of concentrated human use like small towns, garbage dumps, public parks, and prisons. Long com-

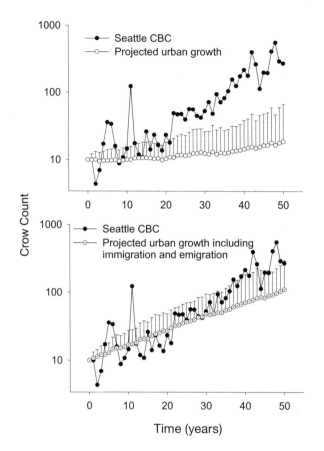

Changes in Seattle populations of American Crows from 1950 (time 0) until 2001 (year 51; after Withey and Marzluff 2005). Solid circles are actual Christmas counts adjusted for annual effort. Open circles are projected population changes considering only the local Seattle population's survivorship and reproduction (A), and immigration from the surrounding suburbs (B).

mutes to scavenge from people are normal for corvids on the fringe of humanity. In Greenland, for example, Common Ravens regularly moved more than 90 miles (150 kilometers) between the few garbage dumps. In contrast, urban crows can fulfill their requirements of food, shelter, and social contact within a small area. This allows them to pack densely into cities.[18]

But why do city crows need so little space? Our observations suggest that it is because humans supply a substantial amount of urban crows' food. There's nothing city crows won't order from the menu. They are fond of pizza crusts, hamburgers, French fries, sweet-and-sour pork, fried chicken, and almost any road-killed animal served up along the highway. More than half of our observations of crows feeding in the city were of them eating garbage, whereas away from people we see them eat garbage less than a third of the time. In wild areas, crows catch emerging aquatic insects, pick berries, rob the nests of other songbirds, and catch small mammals, fish, snakes, and frogs. Although these natural foods are locally and seasonally abundant, they appear infrequently and often vary in quantity and quality. Nature's unpredictability means that wildland crows often commute to seasonally available foods and therefore need large territories. Insects and earthworms are also important foods for city crows, especially during the breeding season. The city birds, however, can also turn to a regularly stocked garbage can or dumpster, thus requiring a smaller territory than their wildland counterparts.

The importance of extensive lawns to crow populations is clear when we see how crows first move into different sorts of neighborhoods. John Marzluff and his graduate students counted crows at the same spots during each of the first five years of forest clearing and house building at five new neighborhoods near Seattle. Crows immediately explored these new subdivisions, but their numbers were only slightly greater than those at nearby forested reserves until the fourth year of construction. At this point, crow populations increased more than 300 percent in large developments that replaced once-extensive forest with grass, roads, community parks, and houses. In contrast, at two developments where homeowners occupied large lots with small lawns and acres of forest, crow abundance actually declined to its lowest point and was virtually equal to that found in uninhabited forest reserves. Apparently, the initial explorers of such low-density developments were unable to find food sufficient for breeding.

Urban crows that are junk-food junkies may pay a price in terms of reproductive success for their subsidized diet. In general, crow reproduc-

American Crows dine on urban refuse

tive success peaks at intermediate to low levels of settlement. It is lowest far from people, but it is also low in our densely settled urban business districts. In the Seattle area, suburban crows average nearly one more fledgling per nesting attempt than do city-center crows. On the nearby lightly settled Olympic Peninsula, crows fledge twice as many young if they live within six-tenths of a mile (one kilometer) of people than if they live more than three miles (five kilometers) from us. Crows studied by Carolee Caffrey in urban Los Angeles have low reproductive success. New York crows also fare best in rural, rather than highly developed, areas. Kevin McGowan of Cornell University has closely monitored crow reproduction in New York for more than a decade and reports that reduced access to earthworms by crows in Ithaca is likely responsible for their reduced reproductive success. Crows in urban Seattle also have little access to earthworms, relative to their suburban neighbors. But the advantages of suburban living around Seattle are also accentuated by active feeding of crows by people. We demonstrated this by supplementing a pair of crows that visited Marzluff's feeder during three consecutive breeding seasons. Each year they laid two more eggs and fledged two more young than their unsupplemented neighbors. They did this year in and year out without regard to climatic variations that affected productivity of other nearby crows. Because the growth rate of crow chicks and perhaps even the resulting adult body size of crows can be increased by abundant food, these well-provisioned suburban youngsters may actually have an additional benefit in being able to survive lean times and compete successfully for limited mates or breeding territories. Beyond the obvious connection between crow reproduction and food, it is also possible that crows breeding in densely urbanized areas might be subject to some of the same pollution that affects human health and reproduction.

Our greatest effect on crows is on their survivorship, not on their reproduction. Simply producing more young crows from nests does little to increase crow populations because subsequent mortality reduces their numbers. Increasing the survival of breeders affords more chances to reproduce and is a more certain way to increase populations of long-lived animals like crows. American people seem to be doing that. Less than one out of every ten breeding crows we followed in our Washington study died in any year. This near immortality was even greater than the survivorship of Common Ravens on Washington's Olympic Peninsula, where one out of every four breeding ravens died annually. In theory, raven mortality should be lower than crows because mortality rates typically decline with increasing body size in birds. Moreover, on the Olympic Peninsula, living with people is what matters to crows. There, if an adult crow lives near people, it is likely to survive, but if it lives more than three miles (five kilometers) from people, it will likely die. Mortality over a two-year period was 2.3 percent near people and 38.9 percent far from people, which gives a long potential lifespan. Maximum recorded lifespans of crows in nature are fourteen to sixteen years, but this will certainly change as studies of banded birds continue. In fact, if only 5 percent of adult crows die each year and we assume moderate survival of young crows, then for every one hundred crows born in your neighborhood, at least five should live longer than forty years.[19]

Our knowledge of crow reproduction and survivorship suggests the existence of fundamental ecological differences among crow populations that are influenced by people. Populations in remote wildlands are not likely to be self-sustaining. They persist only because crows unable to secure land in settled or recreational areas disperse into wildlands as a last resort. Populations there may even go extinct for a few years. This pattern may have been normal early in the European settlement of North America—crows dispersing from abundant populations near people tested waters further afield but often failed to colonize them until people moved closer and transformed the local landscape. Crow populations in lightly settled areas, small towns, and suburbs, in contrast, are extremely productive. There, adult crows produce more than enough young to replace themselves. As a result, local populations grow and expand the species's range if nearby, historically unsuitable habitat is favorably transformed. Dispersers from these productive populations may also help establish and reestablish wildland populations. Eventually, productive populations may become dense as distant habitats become increasingly hostile, territory size may decrease, and reproduction may just balance mortality. These types of stable, self-sustaining populations are seen today in our most urban landscapes.

If urban crow populations are simply self-sustaining, why are so many exploding in size? Immigration is the answer, we suggest. Just as the global human population is becoming more concentrated in urban areas because of the availability of jobs, so too young crows are moving to the cities to exploit their riches. In Seattle we routinely see large collections of apparently nonterritorial crows hanging out at ferry docks, burger stands, and parking lots. Many of these birds, we hypothesize, are the product of successful reproduction along the urbanizing fringe of Seattle or in less densely settled neighborhoods a few miles (kilometers) from the central business district. Juveniles from these productive areas disperse for up to several years to the city, where they are protected from human persecution and can rely on steady grub. In the city, meals are often superabundant and are regularly replenished. Check the dumpster at your local restaurant or supermarket some

Two neighboring male American Crows meet at their territory border as their mates forage in the lawn for worms

American Crow with radio-transmitter. A harness of Teflon ribbon holds the transmitter in place on the crow's back.

day, and you'll understand. Abundant food like this makes the picking easy and less likely to be contested by dominant adults, which likely find it unprofitable to chase and fight hordes of young crows.[20]

To test this notion, John Withey, a well-traveled Seattlite, returned home, entered graduate school at the University of Washington's College of Forest Resources, and helped Marzluff capture and radio-tag fifty-six recently fledged juvenile crows. We followed them closely over two years to determine if those born in productive suburbs dispersed to the city and contributed to the exponentially increasing midwinter counts. More than a fourth of the juveniles dispersed from their home territories during their first year. Nearly half of the twelve that left their homes in outlying suburbs and wildlands moved to more urban areas during their first winter, an average distance of 5.6 miles (9 kilometers). In addition, two of the three that dispersed from the city remained in urban areas. Females were more likely to leave the suburbs for the city than were males. Our conservative estimate suggests that over 70 percent of the annual increase in Seattle city crow counts is being fueled by excessive reproduction in, and subsequent dispersal from, the suburbs (see the illustration on page 97). We suspect that the rest of the increase may come from pockets of productive crows within the city and from longer-distance movements throughout the region. In either case, it appears that cities like Seattle may beckon as a residential paradise for nonbreeding crows from vast areas. These nonbreeders cause the abundance of crows in cities to rise rapidly while the number of breeders remains stable and fully saturates all suitable habitats.[21]

Natural populations do not increase indefinitely. Rather, as English economist Thomas Malthus argued two hundred years ago, growing populations eventually deplete a limiting resource that halts their growth. Malthus was referring to food as he argued that starvation would limit human numbers. The populations of a wild species may eventually level off at a size sustainable by local resources, known as a "carrying capacity." In urban areas, however, carrying capacity may increase faster than animal populations. Food may never become limiting to crows, and as new suburbs are carved daily around many cities, breeding space may also not be limited for many generations. Americans are creating acres of new crow habitat each day as forests are cleared and swamps drained for houses and their support systems of shopping malls and restaurants. In the Seattle area, for example, 204,108 acres (826 square kilometers) of forest unsuitable to crows was transformed into forested urban areas and lawns—prime crow habitat—from 1991 to 1999. In suburbs there is usually a single breeding pair of crows per acre (3.2 per square kilometer). Therefore, in Seattle alone we have created new habitat for more than twenty-six hundred new breeding pairs in only eight years. Eventually, our subdivisions may merge, creation of new habitat may subside, and crows will fill the land at some huge continental carrying capacity. If this scenario plays out across the nation it may take centuries for the American Crow population to reach its new, expanded carrying capacity. Most of the continent would be crawling with crows. Every second or third home property would be defended by a breeding pair and a few helpers. Rich food areas in the commercial-industrial portions of our cities would accumulate twenty to a hundred nonbreeding crows. Landfills and agricultural fields would be black with crows.[22]

Could natural events change this forecast? Yes, the social habits of crows make them vulnerable to epidemics. Currently, West Nile virus is raging through the United States, killing many crows in its path (see chapter 7). Diseases such as this are the most likely natural limiting factor for crows. One can imagine, too, that in specific locations, a predator, such as the Red-tailed Hawk (*Buteo jamaicensis*), Barred Owl (*Strix varia*), Raccoon (*Procyon lotor*), or Great Horned Owl (*Bubo virginianus*), may become adept at harvesting crows on a grand scale. In addition, global climate change will certainly affect crows. Where temperatures increase and water remains abundant, crows will prosper. But where temperatures decrease or aridity increases, crows will decline. Increased fluctuation in winter weather may be especially limiting: freak cold blasts have decimated crow populations in the past. Urbanization tends to mellow climatic variation, however, so risk to epidemics remains the most likely limiting factor for crows in the foreseeable future.[23]

While we affect crow population size and distribution, we also can affect the weave of crows' basic social fabric. Consider crow family size. Kevin Mc-Gowan studied more than three hundred crow families in New York and concluded that suburban families were larger than rural ones. The only place McGowan observed crow families with five or more members was in suburban settings. The largest families had up to eight members, including a breeding pair and three to six helpers. If breeding space is continuously created in suburban settings, as it is in the sprawling subdivisions around Seattle, crows may opt out of helping and actually breed during their second year of life. Suburbia's reduced numbers of predators and abundant food may favor these responses, but cultural evolution may let crows quickly adjust social behavior to their environment. Crows often remain in environments similar to those in which they were born. Young crows learn behavioral patterns from their parents. Clearly, our domestic activity and lifestyle affects theirs.[24]

The ability of crows to exploit the by-products of our lifestyle means that we are responsible for their success. We buffer crows from most of the vagaries nature can throw at them. Urban food is predictable and abundant, their natural predators are not encouraged, and city climates are relatively mild. Unless we change, crows will endure as common compatriots of future generations of Americans and continue to inspire, challenge, compete with, and irritate our descendents. As long as we convert remaining native forests, wetlands, and fields to safe suburban havens that are well stocked with easily obtained refuse and well-watered worm gardens, crows will thrive and mirror our sprawling and urbanizing ways.

Reflection of an urban world

FOUR

Inspiration for Legend, Literature, Art, and Language

"[The raven] is a bird whose historical and literary
pre-eminence is unapproached."
—R. Bosworth Smith, *Bird Life and Bird Lore* (1905)

The clacking of the long cedar beaks echoed like gunshots in the semidarkness. All eyes were on the three native dancers who snaked in and out of a traditional longhouse wearing full raven costumes. Their heads supported brightly colored masks of cedar with strips of bark for hair and four-foot-long black bills. They moved gracefully among rocks and trees looking this way and that like curious birds surveying for food, information, or both. Members of the Kwakiutl (Kwakwaka'wakw) Raven Clan had allowed these dancers to perform the *hamatsa*, or cannibal dance. Each

Kwakiutl raven mask

dancer represented one of three cannibal birds, Raven, How How, and Crooked Beak. Crooked Beak has a curled, deformed bill, but the other two exude sheer power. Raven's face is obvious in all three masks. This ancient ritual portrays the Kwakiutl Nation's creator as mischievous, tricky, variable, and strong. Part bird, part person, these ravens commanded our attention, showing us how these native people represented and revered Raven. Then suddenly we were fast-forwarded back to reality as the lights brightened and the audience began to clap. We had just witnessed a re-creation of an ancient rite on Blake Island in Puget Sound, not ten miles from modern Seattle. The dancers, now clad in shorts, floppy tennis shoes, and tank tops, joined us to ride the ferry back to our modern "village." Crows, not ravens, gathered to perch on posts and greet us at the dock. The young dancers chatted with friends, their radios playing rock music. Although dancing like a raven was artistic and profitable, it no longer seemed integral to these Native Americans' spiritual life. In the company of Raven's children we had arrived at the world of steel and glass skyscrapers, busy streets, and sprawling neighborhoods. That short one-hour ferry ride had transported us through thousands of years from Raven's World to Crow's World, a vast gulf in which both birds have had an outsized influence on the human spirit and our cultural evolution.

The importance of ravens to our culture, as indicated by the epigraph to this chapter, began early. Cave drawings by Paleolithic peoples some fifteen to thirty thousand years ago in France and Spain include images of ravens. Some drawings are of symbolic ravens on posts surrounding burial sites. Perhaps these early cave people were already worshipping ravens as messengers and guides to the afterlife.[1]

Cave art and early legends do not always clearly separate crows from ravens. Legends of early people credit the same miraculous deeds to both birds. The Inuit, for example, tell how Crow tricked a great chief out of a daylight ball and flew high into the sky with it to illuminate and warm the far north for his people. The Tlingits tell of Raven achieving similar deeds. Likewise, although ravens have been the obvious subjects of native artisans for centuries, many sculptural and two-dimensional images of generalized corvids could very well be interpreted as crows, not ravens. A deeper look at Crow and Raven illustrates the diverse ways corvids affect our culture and shows how people of the twentieth and twenty-first centuries are influenced more by crows than by ravens.[2]

Ravens figure prominently in the folklore and religion of northern peoples around the world. The Norse God Odin was accompanied by two ravens, Hugin (thought) and Munin (memory), who surveyed the world during the day and reported back to Odin each night. Northern Native Americans, such as the Koyukon of interior Alaska, believe that Dotson'sa, the Great Raven, created and re-created the world. The second creation came after a large flood that killed all the animals except those Dotson'sa had placed in pairs on a raft. Ravens are also important in more recent flood re-creation stories such as the Christian version in Genesis, in which Noah first released a raven to find dry land during the great flood. The raven appears to have had other business, for it failed to return, so Noah resorted to his second choice, a dove, which proved more obedient and came back with evidence of land. Jewish folklore tells us that the raven's reputation had already been sullied in Jehovah's eyes because of its repeated violations of a decree against love-making.[3]

Ravens were important inspirational and navigational tools to early Europeans. Warriors used the birds' association with death to intimidate their enemies. Topping a Celtic battle helmet from Romania is an iron raven whose wings flapped as the warrior entered battle. Warriors also used ravens to practical ends. Much as Noah supposedly released a raven to find land, the Vikings located Iceland with the aid of ravens. The mariners of Ceylon routinely carried ravens on their ships to set course for land while sailing around the Indian subcontinent.[4]

One of the best known legends about ravens comes from seventeenth-century England. Despite building hatred of scavengers following the London fire of 1666, legend has it that six ravens were spared from persecution and allowed to live in the Tower of London. King Charles II had the birds brought to the tower after the royal astronomer, Sir John Flamstead, told him that if the ravens were killed, his kingdom would fall. A corps of raven masters was established to care for the birds. Not wishing to tempt fate, British rulers have kept ravens in the Tower of London ever since. When only one raven survived the assault of World War II, Sir Winston Churchill is reported to have ordered that more be brought in. Today's Tower ravens have clipped feathers on one wing to keep them from flying off and receive the finest care by the royal raven master and his assistants.

The coevolution of London's ravens and Londoners' legends may not be as ancient as it first appears, however. Boria Sax has researched the history of ravens at the Tower of London and made a startling discovery. There is no evidence that ravens were ever kept in the Tower before the late 1800s. The legend of Charles II and the ravens appears to be a myth, and a recent one at that. Sax argues that the legend of ravens protecting the kingdom was developed at the end of World War II, perhaps to buoy the spirits of shell-shocked Londoners, provide a living connection to ancient times, boost tourism, or to simply increase job security and esteem of the raven masters. Regardless of whether the myth is ancient or recent, the power of ravens to make people believe the impossible and survive difficult times is obvious.[5]

To native Americans of the Pacific Northwest, Raven was not only the creator but also a clown, mischief maker, shape changer, and trickster. To the Haidas he was "Real Chief" and "Great Inventor." The Bella Bella knew Raven as "One Whose Voice Is to Be Obeyed." Raven created the world for his amusement and people were the most amusing of all animals to him. After coaxing us out of a large clamshell and giving us a warm and sunny place to live, Raven made the rivers flow only one way, so as to not make it too easy for us to travel. He even occasionally tricked hunters to rest under a cliff, which then fell on them, so that he could dine on their corpses. It was believed that Raven's greatest trick of all was to give each male animal

The royal raven master at the Tower of London caring for one of the six ravens thought to be necessary to the maintenance of the Crown since the seventeenth-century reign of Charles II. It is more likely, however, that this myth was initiated much more recently by those hoping to increase tourism and add further intrigue to the tower's infamous history.

testicles, so that he might be entertained by the silly games and preoccupations they then constantly engaged in. Raven, renowned for his gluttony, often stole food from other animals and then lied to cover his transgressions. Even the calls of Raven were a source of information. The Kwakiutl would leave the afterbirth of a newborn son for ravens to pick at. It was believed

Tlingit raven panel

that when the boy grew to manhood he would then understand raven language. Such skills would be of great value, for raven cries were said to foretell, among other things, a change in weather, prospects for hunting, an imminent death, or the possibility of an enemy attack.[6]

A deeper look into the culture of the Koyukon Athabaskans helps us understand why raven spirits are less powerful today than in the past. To the Koyukon, every animal has a spiritual legacy passed down from the first world, the Distant Time before re-creation. Only the shamans could call up these spirits, which they often did in times of need. Sickness could be purged from a person, for example, by a shaman dancing and croaking like a raven over him. Modern Koyukon have lost most of their shamans and blame the incursion of ravens into their towns on this loss. They feel that raven spirits now drift like orphans. As one Husila woman from Alaska put it, "Ravens are living as helpless tramps in a place where they do not belong, seeming to care little about their self-respect." A nature with ravens such as this, she noted, was clearly out of balance. Powerful animal spirits were obviously no longer watching over their animal representations. Could this be why modern Koyukon, as well as the coastal people with whom we rode the ferry, pay less attention to Raven today? Without shamans, their spiritual connections to powerful animals like ravens are frayed or broken. Although the transition from Raven's world to Crow's world may be a biological response of an adaptable species to our increasingly urban ways, a cultural shift has accentuated this transition. Animism, guided by shamans who revered nature and the spiritual force of other animals, gave way to a religious life with explicit edicts requiring the subduing of nature and a technology that further separates us from it. Little remains of the life where human and animal spirits intermingled and became one.

Not all the Koyukon have forgotten the power of ravens. Modern ravens —the descendents of Dotson'sa—are still called on for utilitarian services. Ravens scrounging in a sled-dog kennel are forgiven for acting so poorly but then are asked to help the dogs pull strongly. A hunter's string of bad luck can be reversed if he shoots a raven, which then passes its bad luck to other birds so that they become easy prey for the hunter. The hunter apologizes profusely to the raven, telling it that the killing was a mistake. Ravens can also be called on to grant small wishes. A Koyukon hunter may ask a passing raven to "drop a pack down to him." If the raven then tucks one wing

Changing cultures in the Pacific Northwest. The importance of ravens in the past is evident from the totem pole. But the present belongs to the crow.

and rolls over on his back, the hunter will find game. This behavior is in fact a territorial display of adult ravens that is commonly seen when dueling birds chase each other, so we suspect that Koyukon hunters often find their game.[7]

Like Raven, Crow plays a similar if less well known role in other Native American, ancient Asian, and traditional European cultures. The Acoma of central New Mexico say that Crow created the modern world and even saved it from ravaging fires by dipping his wings in water and dropping cooling liquid on a seared earth. This heroic act turned Crow black.[8]

"The Story of Crow," an account of religious heritage by the Tutchone Athabaskan Indian elder Tommy McGinty, suggests that some in the Pacific Northwest did not distinguish between Crow and Raven as creator. This story is strongly held and is placed by the Athabaskans on a par with the Christian creation narrative. "Crow is the one who put up the world—first time. And in the early days, way back, Crow could turn into man and back into bird and fly around—either way, back and forth."

Paralleling the Koyukon and Christian narratives, the Tutchone elder shares the story of a great flood, which clever and industrious Crow survived and created the world anew by piecing together fragments of land. Then, because he wasn't satisfied with the result, Crow summoned a second flood to cleanse the world and start again. As McGinty tells it, "the world we now live on came after that one story."[9]

In some stories, crows even rescued Raven. The Coast Salish of Vancouver Island, Canada, tell of Raven stealing berries from industrious crows and then blaming the loss on enemy raiders. After seeing the "bloody" Raven, the crows began to nurse him back to health. When a small snail told the crows that Raven was the real thief, they became angry. They noticed that the "blood" on Raven was really berry juice and began to mob him and box his ears. The crows made Raven row their canoe home and explain why there would be no blackberries for dinner. It is easy to see what motivated the Salish people to create this story: in nature crows routinely mob ravens to keep them from raiding their nests or stealing their food.

The Tseshaht people of Canada also tell of Crow helping Raven. In their legend, Raven tricked Son of Deer into walking deep into the forest, where Raven then ate him whole in one big bite. The feast was too much for gluttonous Raven, who, back in his house, moaned loudly, fearing death. Crow responded to his cries, and men carried Raven, with his sleeping robe wrapped tightly around his stomach, out of his house. As Crow removed the robe to see what was the matter, she saw the shape of two small antlers pushing out of Raven's stomach. Crow scolded Raven for being so greedy, and the men cut open Raven's belly to free the young deer, which was still alive. Bleeding and alone, Raven was nursed back to health by Crow.[10]

Crows bear gifts in more modern versions of Native American stories. The Hopis tell their children that "the crow has on a blue moccasin" when kachinas bearing gifts are approaching. Kachinas are supernatural beings believed to be the ancestors of living humans. Similarly, Acoma children scrutinize arriving crows to see if they are carrying presents from the kachinas.[11]

Pueblo Indians made connections between crows, death, and bad luck. Their witches could transform into either crows or owls. The Tewas had a war fetish made of smooth black stone and long, black crow or eagle feathers. Fetishes are objects with supernatural power. Tewa warrior priests carried this fetish into battle and used it to behead the enemy. Not all crow plumage brings power. Crow feathers are taboo on corn-mother fetishes, for example, because crows eat dead things. The inconsistent view of crows is exemplified in crows' relationship to life-giving rains. On one hand, the crows' arrival in late summer is celebrated because it is associated with late summer rains, but if the rains fail to materialize, then the crows are blamed.[12]

Native American peoples believed that crows molded other animals as well. We are told that an unsuccessful fisherman, destined to become a crow, cut out his partner's tongue and stole his catch. His partner became a loon, which is why the loon's cry is so mournful to this day. The Lakota Sioux believed that crows warned other animals about approaching hunters. This is why, they explained, crows are now black. Originally white, the crows' leader was tricked and captured by a Lakota hunter disguised as a buffalo. As punishment for continually spoiling hunts, the Lakota hunters threw the crow into the campfire. Crow narrowly escaped with his life but not with his white color. Ever since then, all crows have been black.[13]

Competition between crows and native peoples must have been fierce. The Zunis tell of crows stealing their corn. A Sioux legend has it that one village was so ravaged by crows stealing jerked buffalo meat that the chief called for all crows, their nests, and their eggs to be destroyed. The youngest surviving crow was brought to the chief and raised as a pet. The crow learned to speak many languages and acted as a spy for the chief, warning him of all impending danger. The village prospered because others dared not attack without the element of surprise. The pet crow heard news one day of the chief's death: a bolt of lighting would kill the chief and his brother. Heartbroken, the crow told the chief, thus allowing him to face death in full ceremonial attire. After the chief's death, the crow sang pitifully and flew off. No crow ever visited that village again.[14]

Crows seem to engender legend wherever they live near people. In Jamaica, Man-Crow ruled the island forests, casting darkness over the land whenever he spread his huge wings. The king offered money and his daughter's hand in marriage to anyone who could kill Man-Crow and return light to the world. Thousands tried, but only one, a boy named Soliday, succeeded. Soliday began to sing to Man-Crow, and each time, Man-Crow sang back and came a bit closer. Finally Soliday lured Man-Crow close enough to shoot him with an arrow. Anancy, the Jamaican trickster, tried to claim credit for the good deed but was foiled when Soliday presented the king Man-Crow's huge corvid head. For his treachery, Anancy was cast to live among the rafters as a spider. Soliday got the girl and the money, and the island paradise got back its warm sun. The duet between boy and beast in this fable surely builds on the natural curiosity of Jamaican Crows and the tendency of crows to call back to good mimics.[15]

The influence of ravens rather than crows on our cultural roots is more pronounced because many early peoples lived on coasts, on tundra, and in northern forests where these birds were common. Many of their cultural artifacts have persisted, emblazoned with raven figures. Linking early human cultures unequivocally to crows is hard because early ethnographers may have mistaken "crow" for "raven." However, depictions of crows and ravens interacting, and references to corvids from places like Jamaica that have no ravens, suggest that where people lived with crows, they incorporated them into their stories, legends, and art.

The difficulty in linking early cultures to crows was exemplified in the summer of 2002, when the Marzluff family traveled to Crow Agency, Montana, to learn more about the Crow people. Crow Agency is the geopolitical heart of the Crow Nation. The Crows are a rugged Native American people that once thrived on the cold, windswept plains of present-day eastern Montana, Wyoming, and southern Canada. Their smooth faces and sharp features seem almost to have been sculpted by centuries of wind. Like many Plains Indians, the Crows were nomadic hunters whose lives and welfare were dependent on the great buffalo herds. With the Indian wars of the nineteenth century and the wanton slaughter of the buffalo, the Crow people suffered greatly.

Oddly, Crow legends are ripe with the importance of buffalo and eagle, but not crow or raven. In a moment that taught us more about Euro-Americans' domination of the continent than hundreds of books and films ever could, a Crow woman said simply that her people's English name was a mistake. Her ancestors, she explained, told early explorers that they were known as the "people of the large-beaked bird." In error they interpreted the description to mean "crow," and with such arrogance a people were named. The woman we spoke with said that "eagle" was the correct bird. Others have suggested that the raven was meant. Certainly ravens would have been more common than crows a century ago in eastern Montana, but the woman we spoke with and the beadwork, carvings, and drawings at Crow Agency clearly pointed to eagle, not raven, as being the "large-beaked bird." Regardless, being a "crow" was inconsequential to the woman we met. She was a dignified Absaroka.[16]

On a sunny afternoon in Bothell, Washington, a glimpse of an old sedan reminded us of the influence of corvids on literature. Its license plate read, "THRAVEN," and the plate frame had "Quoth" on top and "Nevermore" below. Edgar Allan Poe, of course, penned the famous line "Quoth the Raven, 'Nevermore,'" in his poem about the dark trickster whose eyes had "all the seeming of a demon's that is dreaming." Charles Dickens, himself a proud owner of a pet raven, incorporated the trials of Grip, the devil raven, in his novel *Barnaby Rudge*. Lewis Carroll put crows on the other

The poet Edgar Allan Poe confronts his raven

side of the Looking-glass. To Alice's delight, "a monstrous crow, as black as a tar-barrel," broke up a fight between Tweedledum and Tweedledee so that Alice could get directions out of a forest in Wonderland.

Crows provide many important lessons for life in the ancient Greek tales known as Aesop's fables. Crow and Raven teach us that fair-weather friends are not worth much, that it is better to be quietly content than loudly defensive, and that we should be content with our appearance. Crow is jealous of Raven in one fable because people see Raven's call as a good omen but Crow's cry as inconsequential. Even though he calls loudly to passing people, Crow attracts only fleeting attention, teaching the reader that assuming someone else's character is ineffective and foolish.[17]

Long before the Greeks, the Sanskrit collection of beast fables known as the *Pañca-tantra* ("Five Chapters") featured crows in informative allegories. In these tales, attributed to a learned Brahman named Vishnusarman, an understanding of crows and owls is used to convey instructions to the three dull-witted sons of a king.[18]

The tales have a simple premise: the King of Crows, Cloud Hue, seeks advice from his advisers on how to overcome his enemy, Foe Crusher, the King of Owls. Within this context the narrator advances a wealth of advice on matters of waging war and overcoming one's enemies. The five crow advisers, wise and opportunistic, offer Cloud Hue a variety of strategies for dealing with their owl adversary. They explore in fine detail the strategic arts of conciliation, confrontation, withdrawal, diplomacy, and deception.

The early Roman naturalist Pliny the Elder also used the natural abilities of corvids as teaching examples in his work *Natural History*. Pliny was impressed by the raven's cunning, ability to speak, and problem-solving skills. He is apparently the first to transmit the story of a raven that, on finding a bucket of water in time of drought, dropped stones into the bucket, thereby raising the cool liquid high enough to drink. A poster from the U.S. government's Thrift Savings Plan provides a modern rendition of this tale, claiming that saving little by little yields a lot.[19]

Millions of children through the ages have learned right from wrong by reading about Aesop's crows and ravens. Today children have hundreds of books about real and legendary crows to capture their imagination. These stories often point out interesting behaviors and adventures of wild animals. Occasionally, they reinforce how much we have to learn about the corvids. Take *Sammy, the Crow Who Remembered* (1969), for example. This engaging story, now out of print, is about Sammy's ability to remember the people who raised him long after he has rejoined wild brethren. But the photographs of Sammy show that he is actually a raven, not a crow.[20]

Crows and ravens have inspired many American poets in addition to Poe. To Secretary of State John Hay, crows were a simple, enduring sight that reminded him of his childhood in Indiana, heralded the arrival of spring, and graced the busy urban sky. These lines from his poem "The Crows at Washington" vividly portray crows coming to roost:

> The dim, deep air, the level ray
> Of dying sunlight on their plumes,
> Give them a beauty not their own;
> Their hoarse notes fail and faint away;
> A rustling murmur floating down
> Blends sweetly with the thickening glooms;
> They touch with grace the fading day,
> Slow flying over Washington[21]

And contemporary poet Doug Anderson weaves legend and biology together in this poem about crows:

> Crows
> Hunch in the trees
> to gossip
> about God and his inexorable
> experimenting,
> about deer guts and fish so stupid
> you could sell them air

and how out in the deserts

there's a dog called coyote

with their mind

but no wings.

Crow with Iroquois hair.

Crow with a wisecrack

for everybody,

Crow with his beak

thrust through a bun,

the paper still clinging.

Crow in a midnight blue suit

standing in front of a judge:

Your Honor, I didn't

kill him,

just ate him

and I wasn't impressed.[22]

Crows have been believed to provide powerful omens of future events since at least the middle of the sixth century in Asia. The art and science of "crow augury," or the practice of divining from crow calls, got its start in India but flourished in Tibet. Reading the ninth-century Sanskrit text *Kakajarita, or Investigating the Cries of Crows,* immediately impresses on the reader the acute awareness early people had of the attitudes, behaviors, responses, and movements of crows. They clearly understood the functions of crows' vocal and physical behavior because they devised a complex scheme to relate the timing and direction of a crow sighting to coming events. For example, if a crow appeared between six and nine in the morning from the west, a great wind would rise, but if it came from the southeast, an enemy would approach. Scattered property would be found if the early crow came from the north, and a woman would come if the crow flew in from the northeast. In "Crow Augury," William Cassidy catalogs how Tibetan tradition used a crow's direction and timing to determine some very important events:[23]

> 9–noon in the southwest, for this *brings numerous offspring;*
> 9–noon directly overhead, for this guarantees *fulfillment of wishes;*
> sunset in the southeast, for this signals *a treasure will come to them;* or
> sunset directly overhead, so they will *obtain the advantages they*
> > *hoped for.*

Or more simply, that they will see a crow:

> on their right as this bodes *good journey;* or
> eating dirty food as this suggests *food and drink are about to come.*

Not all crow sightings foretell events that are to be happy ones; danger precedes a crow from:

> 6–9 am in the southeast, as *an enemy will approach;*
> 9–noon in the east, because *relatives will come;*
> noon–3 pm in the southeast, for *a battle will arise;* or
> sunset in the south, because *you will die of disease.*

Or more simply, beware of a crow:

flapping his wings and calling, because this signals *a great accident;*
pulling human hair or perching on a skull, because this means *death;*
on a withered tree, which means *no food or drink;* or
with a red thread on your house, because it foretells *a fire.*

A less serious but also much less detailed form of crow augury can be found in an old English counting rhyme about prophetic crows. In *Crows, An Old Rhyme,* the illustrator and author Heidi Holder writes that she heard part of this rhyme from her grandfather and filled in the remainder through research. She notes that the rhyme originally dealt with magpies but was transferred to crows by early English colonists, who on landing on North American shores found no magpies, only crows.[24]

One is for bad news,
Two is for mirth,
Three is a wedding,
Four is a birth,
Five is for riches,
Six is a thief,
Seven is a journey,
Eight is for grief,
Nine is a secret,
Ten is for sorrow,
Eleven is for love,
Twelve is for joy tomorrow.

The centerpiece of Seattle Art Museum's Asian collection is a painting in black ink on a gold background spanning a pair of six-paneled screens and depicting a flock of more than ninety crows. Standing over five feet high, the screens are nearly twenty-five feet long. This impressive work of art was completed in the mid-seventeenth century by an unknown *machi-eshi,* or town painter, during the height of the Edo period in Japan. The highly skilled painter clearly understood his subject, for he convincingly depicts the crows flying, fighting, preening, crouched submissively, strutting aggressively, and calling out to one another. Step forward into the embrace of the screens and you're amid the clamor. Spend time with the work and your mind begins to explore the possibilities of what might be going on with this stirring assemblage of birds.

One explanation of the painting is that the crows represent townspeople rising up to break free from the strict social structure imposed by the ruling warrior class of samurai. The artist, it is argued, has veiled his message by giving the birds the spirited resistance he imagines the human community possessing. The crows' natural animated and demonstrative manner makes this an easy explanation, but far more seems to be going on here than the simple statement of the artist's political stance. At this time in Japan the crow's reputation had not yet been corrupted by Western influence. The indigenous religion, Shinto, included a worshipful approach to nature. The Japanese admired crows for their filial devotion, believing that young crows were well cared for by their parents and that, in turn, they would care for their parents. Crows were considered divine messengers and capable of predicting the birth of a healthy child. Their distinctive call, *ko-ro-ku,* was interpreted to mean "a child will come."[25]

The artist could well have been inspired to portray the beauty and mystery of this gathering of crows for its own sake. Crows were commonly de-

A detail of one of the great seventeenth-century Edo screens of crows

picted on lacquered boxes, fans, and hanging scrolls in Edo Japan. The *Screen of Crows* is an expression of the artist's admiration and awe over what he had witnessed. Rather than expressing sympathy for the working class, the painter was more likely giving expression to a more profound experience—the linking of humankind with nature. The *Screen of Crows* captures the essence of the emotional gathering of crows we witness today on a winter's morning when the local crows hold what appears to be an avian rally before moving out to various foraging areas (see chapter 5). This Edo work of art distills and amplifies the reality of nature and invites us to reflect on the unknown. For a moment, as we gaze on the scene, our supposed preeminent position in the world of life is replaced with humility as we reflect on how much more there is to know and understand.

Another even larger screen of crows is in the collection of the Minnesota Institute of Arts. Many of the birds depicted here are identical in posture to the birds found in the Seattle work, and it is likely that the same artist painted both monumental screens. In this second screen the crows are ascending to an adjacent roost, but the artist achieves the same powerful expression of the crows' frenetic activity. Looking at the screen makes it easy to imagine the din of calling birds and the flapping wings of the flock taking to the sky.

A hundred and fifty years after Edo artists celebrated the crow, American artist John James Audubon treated this bird with the same artistic attention he lavished on his other avian subjects. In Audubon's hands, birds become subjects for artistic interpretation rather than simply species for documentation. A pair of Fish Crows, painted life-size, has one of the birds preening in a catalpa. The American Crow, perched in a walnut tree, has a remarkable posture and facial expression: one senses the bird's careful scrutiny of the observer. Like a living crow, Audubon's crow seems to be studying you, sizing you up as if considering what your next move might be.

Audubon also shared his admiration for the crow in his writing. He expressed regret over the wanton slaughter of forty thousand crows in a season in one state. He credited the crow with devouring "myriads of grubs every day of the year that may lay waste to the farmer's fields; it destroys quadrupeds innumerable, every one of which an enemy to his poultry and his flocks." Were it not for the crow, he continued, "thousands of cornstalks would fall prostrate, in consequence of being cut over close to the ground by the destructive grubs we call cutworms." Audubon's wish that "Americans would be more indulgent toward our poor, humble, harmless and even serviceable bird, the crow," strikes an ironic note with modern readers, for a few sentences later he noted that like the raven, crows are "tolerable food when taken a few days before the period of leaving the nest."[26]

By the end of the nineteenth century the crow image was integrated into American paintings to portray a broader vision of nature's forces at work. Winslow Homer's *Fox Hunt* (1893) is a powerful depiction of a pair of crows harassing a fox that moves desperately over a field of snow. The birds, employed here as harbingers of doom, have marked the animal for a fatal end. In the distance, the crow as a metaphor for death is strengthened as black silhouettes roil up above the horizon against a winter sky and head in the direction of the animals in the foreground. The fox's time has come; there is no escaping these unrelenting creatures.

Following Audubon's initial issues of *Birds of America,* German and English artists began including large hand-colored lithographs and etchings of crows within editions of books illustrating European birds. The superb animal illustrators John Gould, Joseph Wolfe, John Keulemans, Archibald Thorburn, and George Lodge drew Carrion and Hooded Crows. These pictures, while expressively painted, were intended to provide information on the species, defining shape, color, and pattern of plumage. At this time the crow was also re-emerging in European art as a metaphor for emotions and ideas.

In July 1890 Impressionist painter Vincent Van Gogh was working amid the farmlands of Auvers, France. Here he finished *Wheat Field with Crows*. Looking at this canvas you confront the crows of the middle foreground. They seem prepared to follow the rest of the flock that swirls away against a turbulent sky to dissolve in a distant vortex of blues and grays. Before us Van Gogh has placed a narrow road—a way through the middle of the wheat field—that turns gradually to follow the course of the crows. The path beckons.

This would be Van Gogh's last painting. A few months later, not far from this field, he would step behind a barn, put a bullet into his chest, and die shortly thereafter. He once told his brother, Theo, that "the truth is we can only make our pictures speak." Like Homer, Van Gogh employed the crow to represent the inevitability of death. Perhaps he sensed his path to be like the fox's.

With an impressionistic painting style and an emotional attachment to the wilds of his homeland, the Swedish artist Bruno Liljefors began depicting nature on a grand scale in the late nineteenth century. A trained and disciplined artist, Liljefors chose to paint birds and mammals not for illustrative purposes but as artistic statements alone. Crows became attractive subjects in his paintings. Bold in pattern and stark of color, they provided Liljefors with forms that not only brought weight and vitality to his design but carried an emotional and intellectual message. His large canvases designed his life-sized subjects into the fabric of the surrounding fields, woods, and waters of northern Sweden. Hooded Crows, with their well-defined black and gray patterning, show up frequently in his pictures. These are animated birds, whether mobbing a hawk or lounging on a split rail fence on a sunny day. In Liljefors's dynamic but orderly system of nature, the crow was a central figure.

Crows still do not enjoy the popular artistic depiction of more colorful and benign species. And yet, as the naturalist painters and sculptors begin to experiment more with corvids, they have provided some bold possibilities. In 1980, American wildlife artist Don Eckelberry painted a particularly memorable image of three crows flying out over a field in autumn. This untitled painting creates a compelling pattern of strong black forms against a muted background. Without the distraction of details, it's easy to become caught up in the crows' flight to a distant woodland. We can attach our knowledge of crows to Eckelberry's birds, and we look beyond them into the painting wondering where they're headed. The social compact that characterizes the species is strongly evoked in the painting's design, and we consider what collective voice they are responding to.

A crow carved of black marble and glistening with a patina created from the continuous rubbing of many hands stands in the town hall of the City of Lake Forest Park outside Seattle. Like its living counterpart outside the building, the sculpture seems to size up the visitors as they pass by. The piece, by Tony Angell, was commissioned by the city to recognize one of the community's more talkative members. Entitled *Town Crier,* the sculpture has now been incorporated as part of the city's logo and newsletter.

American popular culture often draws on the features, fables, and legends of crows. Do you remember the rich licorice flavor of those chewy, gummy candies called "crows"? They were blacker than any crow could hope to be. Maybe your pond has a Karasugoi in it. This predominantly black koi's name means "crow fish." If you are an Australian Rules football fan, maybe you cheer for the Philadelphia, Adelaide, Los Angeles, or Austin Crows. The defensive, fearless, and rowdy character of the crow apparently represents this sport well. Never quiet, the crow's passion for noise may have inspired such modern rock bands as the Crying Crows, the Black Crowes, and Counting Crows. Looking at the softer side of crows, you will find that the romance of a view from the "crow's nest" has inspired the naming of lodges from Tasmania to Maine.

Our association of crows with autumn gatherings, harvests, and death appears to have combined to put corvids in the center of fall decorations and Halloween celebrations. You can buy plastic crows for your porch, a wreath of "happy crows" for your door, and all sorts of folk art crows for your windows, mantles, and gardens. The Atlantic Spice Company of Rockland, Maine, built on the harvest theme with their line of Three Crow Brand spices and coffee. Ravenswood winery of Sonoma, California, weaves three ravens into its logo and features an animated raven that invites wine connoisseurs to discover their homepage.

Crows have been employed as supernatural forces in film. A movie of 1994 starring Brandon Lee and based on the "goth" comic *The Crow* by James O'Barr, depicted a man brought back from the dead who was a messenger from the beyond. Sadly, Lee died accidentally during the making of the film, adding yet another mysterious arrow to the crow's quiver. *A Murder of Crows* was a fateful movie about a lawyer who plagiarized a book only to have the fame of its authorship slip into an accusation of murder. Alfred Hitchcock's 1960s film *The Birds* featured crows and ravens attacking people and to this day leaves film watchers with an apprehensive feeling regarding these species.

Aesop's fables may have inspired the creation of several other corvid characters. Heckle and Jeckle were twin magpies who never failed to get into trouble. "The Fox and the Crow" cartoon starred a cunning crow who constantly chased, annoyed, and harassed a snobbish fox. Wilhelm Busch's *Hans Huckebein* (1867) has entertained German children for more than a century and gives modern readers insight into earlier attitudes toward ravens. In this story a young boy, Fritz, desires "as any boy would" to have a pet raven. Fritz eventually catches a raven and brings it home. Mischief and danger abound as the raven destroys clean clothes, cooked goods, and the nose of Fritz's aunt. As any hope of getting the marauding bird out of the house is lost, the raven discovers liquor. The raven, now quite drunk, becomes tangled in the aunt's needlework, falls, and hangs until dead. The raven teaches children that there are consequences for their mischief.[27]

Old crows appear to have a special place in our hearts. Their name has been adopted for a smooth Kentucky bourbon, a small town in the Yukon, and an unusual group of Strategic Air Command radar operators, the Association of Old Crows, technicians who interfered with the enemy's radar. The original radar jammers and their equipment were known by the code name "Raven," but these operators apparently embraced crows more than ravens. They called themselves "Crows" because the cacophony of background noise and interference they produced overwhelmed the enemy's detection devices. The ranks of Crow grew from the group's origin in World War II through later conflicts, ending in Vietnam. Today they publish their technical ideas in "Crow Caws," have chapters worldwide, and meet regularly. Most crows are aging now, which likely led to the organization's current name. As they freely admit, "Once you are a Crow, it is only a matter of time before you are an Old Crow."[28]

As we have noted, humans have long associated crows and ravens with death. The common practice of scavenging by these species no doubt spurred such associations, but we have gone well beyond a naturalistic interpretation of flesh eating. Crows and ravens are often viewed as transport-

Crow symbolizing death

ers of the dead, carrying souls to the land of the dead. Sometimes souls were not released. The soul of King Arthur of the Round Table is supposed to survive today as a crow or a raven. Is there more to this belief than simply connecting corvids to scavenging? Our recent experiences suggest that more direct connections between crows and people reinforce the belief that corvids embody the deceased.[29]

Ed Bessetti, a truck driver from Seattle, was loading his truck on a September morning in 1996 when he and his daughter heard a thud and saw a crow on the hood of the truck. The crow looked curious and approached them. Bessetti's eight-year-old daughter offered a cookie, which it calmly took. While the crow ate the cookie, the girl petted the crow's chest and head. The crow did not hesitate even when a second young girl rushed over to join in. Both girls handled the crow as if it were a parrot. After a full twenty minutes, the crow calmly rejoined a flock of twelve others foraging in Bessetti's yard. His mother-in-law arrived shortly after the crow, bearing the bad news that the daughter's grandfather had died during the night. Could this strangely behaving crow really have been carrying the soul of a recently departed loved one, or was it simply someone's former pet? We rationalized the latter, but our beliefs continue to be challenged by similar events. A friend of Bessetti's reported a crow peering in her window for three days after her mother died.

Perhaps it is the crow's nature that makes it a likely link with qualities of people we know well. Tony Angell recalls vividly how the spirit of a good friend seemed to manifest itself in a crow. The bird appeared one morning hammering on Tony's roof. This was an unprecedented occurrence, for the local crows generally gave the yard a wide berth, knowing Tony's habit of discouraging all presence of crows when birds were breeding in his woods. Investigating the disturbance, Tony discovered the crow, which immediately flew down to stand on the porch before him. The crow seemed more interested in mischief than food as it dipped its beak into an open can of paint, jumped to Tony's shoulder, and continued to remain about the porch area throughout the day, scattering brushes, pulling on potted plants, and scolding the dog. Again, the next morning, more roof pounding and general havoc about the porch. The crow was awaiting Tony. This time the bird was taken around the yard on Tony's hand and studied closely. The crow seemed fearless, staring back at Tony's intense gaze with equal interest. On entering the house from a back door, Tony headed for the bathroom to wash his hands, but the crow would have none of it and for the first time cawed aggressively and nipped Tony on the shoulder. Turning toward the kitchen, the crow calmed down and then flew to the refrigerator as if awaiting an inspection of the contents. The bird seemed very familiar and comfortable in this part of the house. It was about this time that Tony was struck by the crow's remarkable resemblance to his friend Fred Harvey, who had been gravely ill in Santa Fe, New Mexico. The bird's curious, proud, slightly imperious, yet mischievous and bright manner was a perfect match for Harvey's personality.

Later that day, the crow stepped up onto Tony's hand and the two stood before the window, looking out to the porch where the crow had first appeared. Tony asked, "Freddy?" The crow responded with a fixed gaze, and as bird and man stepped out onto the porch deck, the crow flew off through the trees and was never seen again. On the following day, Tony learned that his friend had died the day the crow arrived. Short of a metaphysical debate over what has occurred here, we are wondering if others have had similar experiences? Why do crows conjure up the images of a kindred soul?

Few animals have influenced our language more than crows and ravens. A survey of the *American Heritage Dictionary* makes our point. "Crow" and "raven" have been incorporated into nine words, among them crowbar, scarecrow, and ravenous. This is more than for such other wild animals as salmon (one word), whale (four words), coyote (none), fox (seven), frog (one), wolf (three), and eagle (one). Only a few domestic species appear to influence our language more than crows and ravens. Our search turned up eleven words incorporating "cat," twelve with "dog," and eight using "cow."[30]

Traveling the British countryside simply overwhelms you with the influence of corvids on language. Place-names echo a corvid past. Great Raveley, Raven Beck, Ravencroft Belt, Raven Scar, Ravenscleugh, Ravenscliffe, Ravens Close, Ravensheugh Crags, Ravensnest Wood, and Ravenglass are just a few. Ravenstone was a European execution site. In Scotland, ravens are known as "corbies," and the Highlands resound with places like Corb Glen, Corbie Head, Corbie Nest, and Corby's Crags. In all, more than four hundred British place-names can be traced to the raven. The propensity with which we name our places after the raven has been so great that biogeographers could accurately map the early distribution of the raven just by mapping raven place-names.[31]

Crows and ravens have profoundly enriched many languages, but humans have been remarkably simple in developing names for corvids. Greg Keyes, an anthropologist at the University of Georgia, has documented 181 names for crows and ravens from 136 human languages on five continents and many islands. Most names are similar and distinctly onomatopoetic. Most include a "k" consonant sound and an "a" vowel. Well over half of the words for crow worldwide include "ka" or "ak." Finns, for example, call ravens "korpi" and Poles call them "krucks." Batanes Island Itbayats call crows "quwaks," Pawnees call them "kaaka," and Thai call them "kaa." Certainly these reflect the cawing, clucking, and croaking calls typical of crows and ravens. Few words in any language are so eloquently derived from the sounds of nature. Even fewer onomatopoeic words survive intact as languages evolve. The consistent and persistent names given to crows and ravens show the power of their voice. It may also indicate our long and close association. As people dispersed from Africa and Europe they must have remembered their native crows and recognized newly encountered species of similar build, demeanor, and voice. They named these new associates as they had the old, using the voice spoken by the beast.[32]

As we have seen, English writers frequently mention the crow and raven in their works. A popular English proverb from the 1500s says simply, "An evil crow, an evil egg." Another from 1640 advises that "the devil . . . sends his black crowe, anger, to plucke out his ey." Britons acquainted with the supernatural were said to possess "raven's knowledge." Shakespeare, never one to overlook a graphic image, employed the crow frequently in his writings. In his great tragedy *Macbeth,* for example, "Light thickens, and the crow makes wing to th' rooky wood." In fact, Shakespeare himself was called an "upstart crow" in 1592, by one Robert Greene, a popular English writer of the day. Convinced that Shakespeare was a plagiarist, Greene drew an analogy between Shakespeare, who used others' words to gain fame, and Aesop's crow, which wore peacock feathers to look more beautiful.[33]

Americans also embellished their version of English with references to the crow. Integrating freed slaves and other black-skinned Americans into the South's dominant white culture was slow. Jim Crow laws actively discriminated against African Americans. The term *Jim Crow* appears to have been coined in 1837 in a poem by R. H. Barham. In the poem Barham refers to a thieving jackdaw that steals a Catholic cardinal's holy ring. The ashamed bird returns the ring, abandons his life of crime, and is canonized Saint Jim Crow. Like Aesop's crow, wearing peacock feathers, Jim Crow lived beyond his rightful social class, gaining more respect than provided by birthright. Another saying that is still used in the South has its origins in sixteenth-century England: "Every crow thinks her own bird is the fairest."[34]

Scarecrow conjures up the familiar image of a human form designed to

Ancient cultural coevolution between farmers and corn-thieving crows. A Mimbres potter from the tenth or eleventh century created this bowl featuring noosed crows hanging from a garden fence. Two uncaught crows, learning by observation, are escaping the farmer's detection.

scare crows and other pests from fields and gardens. Native Americans employed smoke as well as the presence of women and children to keep crows and ravens from their drying fish, but in one picture of a Makah fish-drying rack from the late 1800s, a familiar scarecrow is clearly built into one end of the rack. As early as the sixteenth century, young European men were employed as *crowboys* or *crow-herds* to guard wheat fields. The boys are long gone, and today's stationary scarecrows are more useful as folk art than as repellents, for crows and most other seed-eating birds quickly habituate to

Scarecrow

their lack of movement and ignore them. Efforts to improve their effectiveness have included giving scarecrows moving parts and moving them around the fields. Dead crows are often hung on fences or from trees near fields. This can be effective for months, because crows readily associate death and danger with places where their brethren have died. This modern practice has a long history. A Mimbres Indian bowl from the tenth or eleventh century in New Mexico features three crows hanging from fences. Two approaching crows are shown in the act of learning about a corn-raider's fate. A human holds empty nooses, perhaps testifying to the challenges and persistence of former crow trappers. The ancient potter who so skillfully presented the crows and trapper gives us a glimpse of cultural coevolution in progress. At least a thousand years ago, crows were adjusting their culinary culture to include corn, learning about traps, and stimulating the human cultures of pot-making and crop defense. Coevolution between scarecrows and crows is literally causing a cultural radiation. Modern "scarecrows" are more often air guns that deliver concussive blasts at random intervals to keep even the most confident crow on edge. Scarecrows have evolved out of the agricultural fields and into our art, literature, film, and fable. Boria Sax concludes that today's scarecrows are "ghosts, sorceresses, vampires, dancers, space aliens, rap stars and assorted demons."[35]

"Warner and Clark are 'eating their daily crow' in the paper."
—Mark Twain, letter to W. D. Howells, January 11, 1876

We have all had to "eat crow" at one time or another, although we would rather forget such humbling moments. This phrase is used in a variety of situations to mean having to recant a statement or action that later proved incorrect at best and humiliating at worst. The phrase may have originated at the end of the War of 1812 when a bored American soldier crossed the Niagara River into British territory to hunt. Finding only a crow, he shot it. An alert British officer heard the report and accosted the intruder. Though unarmed, the clever officer asked the American to see his fine weapon. Pride clouded judgment, and the American handed over his rifle to the Briton, who quickly turned the rifle to his advantage. As punishment for trespassing, the British officer made the American take a bite of the crow. Despite his protests, the American ate crow. After escorting the American to the border, the Briton returned the gun, which the American quickly turned against his former captor. Now it was time for the officer to dine, so the Briton also ate crow that day. Had the American shot a grouse rather than a crow, our language might have remained impoverished. Grouse taste great, but crows are presumed to taste like the offal they often consume. That is why "eating crow" could be viewed as severe punishment.[36]

Despite this notion, Inuits, Northwest Native Americans, and Greenlanders often ate crows and even ravens, and some Europeans once dined extensively on young Rooks. Flocks of crows were netted in the nineteenth and early twentieth centuries along European shores. Captured birds were beheaded by specialized "crowbiters," men who "ate crow" for a living. Crow banquets were not uncommon in the United States in the 1920s. Dressed crows could be bought for a mere nineteen cents in Oklahoma, although they would cost you more and go under the name "rook" in Denver— perhaps appealing to the more sophisticated palate. Ernest Good, one of the earliest American graduate students to study crows, often fed them to his family. Apparently graduate students have always sought cheap meals. Good notes that his children were especially fond of crows, preferring them even over fried chicken. Crow feasts are being revived in modern Lithuania. There, crows were a traditional meal from medieval days through the 1930s, but with Soviet domination (1940–1991) crows were left off the tables. Now Lithuanians gather for organized crow hunts, where nests are searched and nestlings are gathered to be "boiled in cooking oil over a bonfire and served with various vegetables," just as was done by medieval nobles. Some crow-meat enthusiasts have even produced a new Lithuanian beer, "Young Raven," to go with their national dish.[37]

Compelled and curious, we two had to try it ourselves. Knowing that "old crow" is better to drink than eat, a fresh roadkill provided us with a young bird, and we headed for the kitchen. We sautéed the crow breast in olive oil, garlic, and red pepper sauce and then fried it. We both agreed the results were tasty, a far cry from your everyday chicken, but we suspect that it will be a long time before Americans are willing to eat crow routinely.

"We cut over the fields. . . . straight as 'the crow flies.'"
—Charles Dickens, *Oliver Twist*, 1838

As the crow flies" has become a mainstay of American lingo. This phrase refers to the straight-line distance between two points and probably comes from observations of crows migrating in spring and fall or traveling to and from communal roosts. At these times, long, straight lines of crows commute directly from point to point. These are exceptional flights, however, because crows are easily distracted during their daily travels and rarely take the shortest route between two points (see the illustration on page 146 for a real-life example). In the Southeast, the term *crow's fly* is used to refer to a place a short distance away.

Crow, in criminal parlance, means one who keeps watch while another steals. Members of a burglary team, called *crows*, assisted by remaining outside as the crime was being committed and kept watch. Should anyone approach, they'd give a signal to alert their cohorts so that all might make

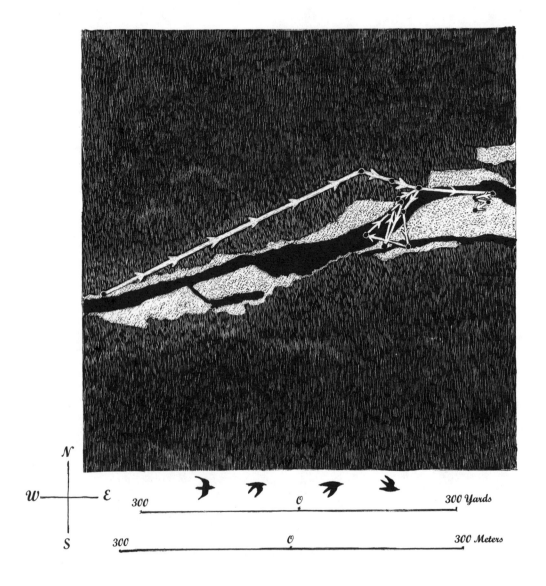

As the real crow flies is not always straight, short, and direct! Despite frequent claims to the contrary, crows don't fly in straight lines all the time, as this sequence of moves by a radio-tagged crow along the Hoh River on Washington's Olympic Peninsula demonstrates. Sequential moves (indicated by dots) during a two-hour period are joined by the line. Direction and sequence of movement are indicated by arrows. This crow made sixteen moves over a total of 2,800 yards (2600 meters) to travel the actual straight-line distance of 740 yards (680 meters).

their escape. "Crow" is apropos for such lookouts, as many flocking corvids are known to post sentinels that will alert their feeding brethren if danger approaches.[38]

"Certeine instruments wherewith they might pull downe the workes y their enemyes made, called Harpagons, and also 'crowes' of iron called Corvi."
—John Brende, *The Historie of Quintus Curtius,* 1553

"Well, Ile breake in: go borrow me a crow."
—Shakespeare, *The Comedy of Errors,* 1590

Even the anatomy of the crow has guided our engineering. To gain leverage, we use a *crowbar,* modeled after the crow's sturdy leg and toes. This is an iron bar usually with one end slightly bent and sharpened to a beak. Used for utilitarian purposes, these *crows,* as they are often called, take a variety of shapes and are employed for everything from levering and prying to drawing nails and driving the crowbar's sharpened point into rubble. This last use led to the crowbar's effectiveness as an agricultural tool for making holes in the ground and as a mining tool for breaking ore loose.

The crow's beak, so efficient for picking up and holding small objects, inspired the design for a pair of seventeenth-century forceps called a *crowbill.* This tool was used for probing into wounds and extracting bullets or other foreign objects. A larger, long-handled version of the crowbill with a special side-claw design was used in mining and extended into deep bore holes to recover broken rods.

The term *crowsfoot* dates back to the seventeenth century and describes a device consisting of a number of small cords rove through a long block that keeps a sail from chafing. It also refers to a stand attached to the end of mess tables and hooked to a beam above. A more obscure use of this term refers to a military device made of iron that has sharp spikes extending so that it always points upward. Used in securing fortifications, this crowsfoot discouraged invading enemies who risked impalement if they stepped or fell on it.

The English navy once used a machine called a *crow* that was equipped with an iron hook and fastened hold of the enemy's vessel to pull it alongside. Peter Weir's film *Master and Commander* (2003), based on Patrick O'Brian's sea novels, vividly showed how British sailors used such crows to secure a French vessel. The land armies of that period used siege

Crow considers its namesake

apparatus that consisted of ladders with *craws,* or clamps of iron, to catch the angles of the opposition's fortifications. Both of these military tools were based on the crow's foot with its long toes and hooked toenails. Of course, a ship's lookout platform, placed near the top of the mast, is appropriately named the *crow's nest.* The name is especially apt, for not only do crows place their stick nests high up in the tree tops, but their remarkable propensity for spotting approaching threats from these retreats is well known by anyone who has examined nesting crows.

"Til crowes feet be growe under your ye."
— Chaucer, *Troilus and Criseyde,* 1380s

"I think I like your horses best. I haven't seen a crow-bait since I've been in town."
—O. Henry, "The Trimmed Lamp," 1907

"At this I . . . rounded up my 'crow bait' and pulled out for home."
—J. Marvin Hunter, *The Trail Drivers of Texas,* c. 1925

Eventually, if we live long enough, we all develop wrinkles about the eyes that radiate outwardly like the outlines of *crow's-feet.* A derogatory slang term for an unattractive older woman is an *old crow,* and once you reach the point of no return physically, particularly if you are a horse, you become *crow bait* or *raven food.* A seventeenth-century writer pushed the idea even further, referring to doomed soldiers as *crows' meats,* while an American southeasterner would probably use *crow bones* to describe such a state.

"He and I very kind, but I every day expect to pull a crow with him about our lodgings."

—Samuel Pepys, *Diary,* November 18, 1662

The prospect of plucking a crow is not particularly attractive, so the expression "to have a crow to pluck or pull" with someone means that you have something disagreeable to settle or clear up with that person.

"As blak he lay as any cole or crowe."

—Chaucer, *The Knight's Tale,* c. 1386–1400

The color of the crow's plumage has been used to describe a variety of subjects. Miners would refer to the color of the ore as *crow* rather than as black. A miner coming upon a poor, thin, or impure bed of coal or limestone would call it *crow coal*.

To *crow-hop* is to hop like a crow and refers to a horse jumping about with arched back and stiffened knees as a precursor to bucking. When you say that someone is *crow-hopping* in America, you mean that he or she is trying to back out of an argument. A *crow hobble* is a rope tied to a fore and back leg of a horse to stop its leaping and bucking. There's no evidence, however, that crow hobbling has ever been used to put a stop to backing out of a dispute.

As early as the sixteenth century, English doctors used the term *crowlynge* or *crowling* to describe the rumbling sounds made in the stomach or bowels of a sick patient. *Crowing* was the term used to refer to the sounds made when the patient had whooping cough or croup.

It may be the ultimate measure of our capacity for disdain of the crow that in New England the word becomes a euphemism for cursing. "Jesus Christ!" becomes "Jesum Crow!" and "For Christ's Sake!" becomes "For Crow Mike!" In Australia a common expression of surprise or disgust is "stone (or stiffen) the crows."

And so it goes, assorted reference to crows being regularly incorporated into the English language. As the crow extends its populations and we expand our appreciation of the crow's behavioral and physical complexities, crows will likely broaden their influence on our vocabulary and provide new figures of speech to enrich our view of the world around us.

We know of no other wild animal that so consistently and thoroughly has affected our art, language, religion, and science—literally since the dawn of human history. There is no doubt that people learned about crows and ravens by listening to their teachers, mentors, and peers, by looking at art, dance, and film, and by reading holy, popular, and scientific writings. That is, we have come to know the crow not merely by contact in nature but by social learning, and we have modified, embellished, and extended this understanding through cultural evolution. As we now begin to explore the behavior of crows, you will learn of the myriad ways we affect crow behavior, but you will notice that our certainty in labeling these changes as arising from cultural evolution is less than our certainty about human cultural evolution. This is often simply because we can never understand another species as well as we understand ourselves.

FIVE

The Social Customs
and Culture of Crows

As night falls, swirling flocks meld and settle among the branches of the large cottonwood trees behind Rene Drake's house. Squadrons of cawing, ebony explorers have been pouring in from the east for more than an hour now, often moving about the tree-covered wetland reserve, a ten-acre (four-hectare) island of open space in a suburban sea. Perhaps five hundred or more American Crows have gathered this August evening to roost, as they have here for the past twenty years. The riot of crow sounds magnifies their numbers. They are insanely loud, with little sequence or rhythm, as if the birds are determined to divulge all their secrets before they sleep. Seeking

A murder of American Crows mobs a Great Horned Owl

respite from the din, Rene occasionally tries to startle the birds by slapping boards together while her neighbor Kevin Grigsby fires an air rifle into the trees. Nothing works. Only total darkness quells the screaming horde.

Frenetically noisy displays like this are a hallmark of crow behavior. These displays catch our attention because they involve hundreds, thousands, or even hundreds of thousands of birds. But what makes crows and ravens more like monkeys, apes, and people and less like other birds are the less obvious social nuances afforded them by their large brains. Crows and ravens care for each other, spend years living at home, engage in foreplay, and mate for life but now and then mate with others. They sunbathe, smear natural oils on their skin, and play. Large groups cooperate to sleep in safety, drive away mutual enemies, and maybe even dole out justice.

As we begin to consider the social behavior of corvids, we need to introduce the idea of *anthropomorphism,* the giving of human qualities to animals. Animal behaviorists, like us, strive to reduce unnecessary anthropomorphism because it can lead to severe misunderstandings about the differences between humans and other animals. Simple actions like sunbathing by people often involve complex thought processes, reference to cultural norms, and hidden motives. But the actions of wild animals can evolve free of such complexities if they increase an individual's survival and reproduction. Natural selection accomplishes this. The social lifestyle and intelligence of crows, however, afford them many opportunities to interact with others and solve problems in ways that at least superficially resemble us. Observers thus often describe the actions of crows in human terms. We do this for engaging convenience, not scientific accuracy. We try to minimize the inherent dangers. Our use of terms like *funeral, foreplay,* and *execution* are intended to better describe crow behavior as it appears generally and are not intended to be literal interpretations that would ascribe the same complex human motives, values, intelligence, and actions. There is a growing body of support in the animal behavioral community for a new way of viewing other species, suggesting that they are capable of using emotions, feelings, and consciousness in a manner similar to our own. We choose not to enter this fray but simply to describe corvid behaviors in a manner that will help nonscientists to understand and evaluate complex behavioral patterns.

Our use of human terms to describe crow behavior builds on the work of Nobel laureate and pioneer corvid behaviorist Konrad Lorenz. Lorenz often described animal actions in human terms, and he clearly related animal behavior to our own behavior. He wrote of jackdaws wedding, playing, and passing on tradition, clearly stating their emotions and intentions. Some of his references may be seen today as mistaken and misleading assignments of human properties to animals, but most show us the potential complexity of animal sociality and illustrate, as Lorenz wrote, "what an enormous animal inheritance remains in man, to this day." If behaviors that increase an animal's survival and reproduction have human parallels, this does not mean that the animal has other associated human traits. It may. But we may only infer that some behaviors can be performed by many species, not only humans.[1]

The crow social life is both fluid and stable, a mix of competition and cooperation. Mated pairs cooperate in nearly every nuance of life, but compete with other pairs for territories. Unmated crows variously compete and cooperate with their parents, siblings, neighbors, and members of large migrating or roosting groups. In crow society, large numbers of individuals interact rarely, but are familiar with each other. Throughout their lives, most individuals have sustained relationships with a few other individuals. Central among these stable relationships is the lifelong, monogamous mated pair. This fundamental unit of crow society defends a territory, ranging from the immediate nest area in Northwestern Crows to hundreds of acres (hectares) in Common Ravens, for their exclusive use throughout their lives.[2]

All crows have some exposure to family life, living with their parents and siblings for at least the first few months after fledging. Some young crows remain with their parents and perhaps up to six siblings and half-siblings for a year or more. However, all crows do not experience long-term family life. Instead, some leave their parents after a few months to disperse widely, wander locally as unmated "floaters," or even form a pair-bond to breed in their second year. Female crows usually disperse and occasionally breed in their second year. Males may wander locally and rarely breed before their third year of life. Whether crows wander or remain in families, they all eventually participate in much larger social gatherings, most commonly during winter, as they forage in flocks or roost in impressive, nighttime aggregations. As Cynthia Simms Parr has noted, this type of ebbing and flowing, dynamic social system is more typical of some primates than it is of birds. In our opinion, this extends to human societies, which may explain our attraction to crows.

The world's crow species differ only slightly in their social details. Although Common Ravens are among the least social, living as mated pairs during their adult lives, they do flock together at rich food resources during the winter and spend their early years in the company of tens to hundreds of other young, albeit nomadic, ravens. Jungle Crows have a similar lifestyle. In contrast, Chihuahuan Ravens, Rooks, Western Jackdaws, Northwestern Crows, and Fish Crows are rarely found outside a flock. American Crows, Australian Ravens, Carrion Crows, and House Crows appear to be more variable, flocking when resources allow it, aggregating to roost each night, and living in permanent, but small, families on well-defended territories most of each year.[3]

The participation of crows in their various relationships varies regionally, seasonally, and throughout an individual's life. Most American Crows from south-central Canada migrate to the southeastern and midwestern United States in winter. Some American Crows in the northern United States also migrate, but many remain on, or at least visit, their territories every day of the year. Crows do not appear to migrate from southern portions of their range or from northwestern areas, including southeast Alaska. It is unlikely that entire families of migratory crows remain together during migration. We suspect that mated pairs remain together, but even local crows that join larger groups during the winter do not remain in obvious family groups. Pairs and families that do not migrate maintain territories and tight social relationships year-round. They may, however, forage independently with local floaters or migrants at locally concentrated food sources each winter day. These flocks do not have consistent membership, but transient relationships among individuals likely develop despite the group's fluidity. In fact, unmated individuals may select their lifelong mates during this time.[4]

Glimpsing details of crow social customs demands perseverance, patience, and luck. We cannot simply conduct an experiment to reveal the rules structuring crow society. Rather, we must watch hundreds of individually recognizable crows live out their lives. Mostly we do this by catching, marking, and releasing wild birds. Occasionally we get the chance to be foster parents, peripheral members of crow society, and thereby we gain truly unique insights. Together with his students Laura Landon and Kate Whitmore, John Marzluff raised hundreds of magpies, American Crows, and Common Ravens. Our mission was to inform restoration efforts on endangered Pacific Island crows by learning how to best collect, transport, incubate, hatch, raise, and release these common mainland corvids that served as surrogates for the endangered species. Along the way we watched seventy-four ravens, fifty-seven crows, and thirty mapgies leave our care and integrate with wild birds. We radio-tagged each bird before release so that we could consistently check up on its socialization. One of our female American Crows let us in on where and when mates are selected. We watched her join and roam widely with a winter flock in southwestern Idaho. She associated with hundreds of crows and eventually recruited a mate from the flock. They returned to breed in a small town near where we let her go the previous year. Earlier surveys of the town never turned up crows, so we knew that the recruited male was not a local. This pair was the first to colonize this growing town.[5]

Seasonally, most crows go from living with their mate and/or family and interacting with a small number of neighbors in spring and summer to interacting with larger groups during fall and winter. The social setting varies with geography and migratory status. Migratory crows from northern regions probably have less stability in their social lives than nonmigratory crows. They probably do not consistently interact with familiar neighbors outside of the four- to five-month-long breeding season. Presumably their lives revolve more around short-term relationships with many individuals than around the long-term family and neighbor relationships that characterize nonmigratory crows. Nonmigratory crows spend portions of nearly every day on their territory with their entire family, usually in and out of contact with well-known neighbors. They do interact with large numbers of individuals for five to six months of each year when migrants or wandering nonbreeders collect at rich food resources around their territories. Nonmigrant territorial families often forage with these vaguely known strangers when food supplies in their territories dwindle. Even if they remain to forage on their territory during a winter's day, they usually join hundreds, even hundreds of thousands, of other crows each winter night to roost communally. In winter they leave their roost mates to survey at least briefly their territories each morning, but in summer, they often roost as a family on their territory. So, migratory crows interact with large numbers of strangers on distant lands each year. Nonmigratory crows also interact with many other crows for half of each year. But they do so on or near their year-round homes in the company of their family, familiar neighbors, and varying numbers of perhaps unfamiliar migrants and vagrants.[6]

S ocial life changes dramatically as crows mature. Some of these changes are gender-specific. Newly fledged crows know only the family social unit for several months, but they quickly join larger foraging and roosting groups, either alone or as a family member, each fall and winter. If they remain with their family during this time, they likely roost and forage with them and return to the group's territory for at least part of every day. Therefore, they may not integrate very far into the larger winter social scene. At some point, however, during a crow's first autumn and winter, it settles on one of at least two options. Some disperse from their home territories to "float" from place to place with other nonbreeders for months to years before they form a pair-bond and obtain a territory. Others remain on their home turf to "help" their parents defend the homeland and rear next year's offspring.[7]

Researchers rarely follow birds to understand where and how these "decisions" are made. Most birds move too far, too fast, and are too small to follow. So ornithologists band (called "ringing" by Europeans) thousands of birds each year in the hopes of seeing some exploring new ground, settling to breed, or remaining near home. Like any gamble, this rarely pays off with more than an occasional win unless many birds are banded and many volunteers help resight them. Fortunately, we have other options for studying large birds like crows. Radio transmitters, less than 3 percent of a crow's body weight and therefore easy for them to carry, will run on a single battery for up to two years. This allows us to catch and tag a reasonable sample of birds and literally dial them in every few days for observation. Young crows move around a lot and we must be within a few miles (kilometers) to receive their radio signals, so we spend hours driving hundreds of miles (kilometers) each day to eavesdrop on our birds. But it pays off.

We introduced John Withey's graduate work earlier as we studied the influence of dispersal on American Crow population growth in Seattle. Those same fifty-six newly fledged American Crows allowed us to better understand the options of juvenile crows. We discovered that young crows are vulnerable; more than half were found dead during their first year. The others, however, illustrated the variety of social options available. More than one-third dispersed to join other crows and wander widely during the winter. One in five remained with their families to help. They assisted by defending the home territory from predators and intruders, and most fed their younger siblings in the nest and shortly thereafter. One male even wandered most of the winter and then returned, in true prodigal son fashion, to rejoin the family and remain on his home territory as a helper. Females were more likely to disperse than males. Every female we followed dispersed, but only 15 percent of males did so. Males were more likely to remain home as helpers than were females. None of the females we tagged helped, whereas nearly 80 percent of males that survived their first year helped.

A male yearling American Crow helps its parents feed nestlings

It is not surprising that one sex disperses more than the other, because this reduces close inbreeding, but why should males disperse less than females? This common attribute of birds may reflect the fact that male crows are often dominant to females or that females invest more in reproduction and are more valuable to a population than males.[8]

Careful tagging and following of American Crows in New York by Cornell University's Kevin McGowan has shed more light on where young crows go to breed. McGowan's crows rarely disperse far from home. He resighted fifty-seven breeding crows that he originally banded as nestlings. On average, they bred within two and a half miles (four kilometers) of their home territory, but a few traveled further; three females born in Ithaca, New York, traveled to different towns to breed. One went thirty-seven miles (sixty kilometers) away. In general, dispersers sought out breeding areas similar to the ones they were born in; rural crows dispersed to other rural areas and suburban crows dispersed to other suburbs.[9]

The helping behavior of young birds has fascinated sociobiologists, scientists who study the evolution of social behavior, for decades because it appears at first glance to contradict Darwin's theory of natural selection. Why in the world would a young bird help another reproduce rather than breeding and rearing its own young? After all, offspring are an important currency of natural selection; those who fail to breed do not pass on their genes and therefore are selected against. Careful study of helping, or "cooperative breeding" as the full suite of behaviors is known, has revealed that helpers often actually increase their genetic representation in future generations by helping. That is, helping is not the altruistic contradiction to natural selection that it first appears to be. Helpers may benefit by garnering extra parental care and all the protections afforded a closely knit social group. This can increase their chances of surviving to breed successfully later. If helpers really help, then they probably increase the number of young their parents can raise, which pays the helper a genetic dividend. Siblings share on average 50 percent of their genes in a monogamous breeding system. Increasing your parents' productivity is not as good as reproducing yourself, but it is far from a total genetic loss. Helpers may also gain experience caring for young that lets them reproduce more successfully later in life, or occasionally mate with a breeding adult, or inherit part of their home territory. McGowan's work in New York suggests that helpers allow a pair to expand their territory, a portion of which is then inherited. This may be especially motivating for young male crows; however, these males may also breed with the resident female. The other benefits may be especially motivating for female helpers.[10]

Crows in New York, Florida, Michigan, Washington, California, and British Columbia have fairly consistent patterns of helping. Helpers are typically juveniles from the previous breeding season. They help their parents feed nestlings and fledglings, provide extra sets of eyes to spot predators, and increase the raucous energy in mobbing and driving off invaders to the home territory. Helping is common, not universal. About 35–95 percent of American Crow pairs in New York, Michigan, Florida, and California had helpers. Northwestern Crows may be less helpful; only 17–20 percent had helpers in British Columbia. Our observations in Washington suggest that about 20 percent of pairs have helpers. Most crow pairs have only one helper, but up to six helpers have been observed. These large families consist of a pair and their young from four or more previous breeding seasons. In New York, suburban crows have greater numbers of helpers than do rural crows. More than a single helper has never been observed in Northwestern Crows. Common Ravens may rarely have helpers. Helpers do not necessarily increase the annual reproductive success of the pair they help. Pairs of American Crows with helpers and pairs without helpers had similar breeding success. A study on Northwestern Crows, however, suggests that, in this species, pairs with helpers may fledge more young than those without helpers.[11]

So if most pairs with helpers do not raise more young than those without helpers, why do parents tolerate helping? Our work in Washington, like studies in Michigan and New York, suggests three important possibilities. First, helping may enhance a breeder's survival by increasing vigilance for predators like Red-tailed Hawks and Great Horned Owls, who gladly eat crow. The secret to lifetime success for crows is to live long lives. This allows them to reproduce and contribute future breeders to the population in each of several years, rather than gamble on a huge brood in just one or a few years. Increased survivorship would therefore do more for a crow's reproductive legacy than would fledging a few more nestlings. Second, helpers may let breeders fledge young more consistently. Rather than fledge a few more young every year, crows with helpers may benefit only in really poor years when predators or low food supplies limit most breeders. The extra eyes to spot predators and food may then pay off. For example, 2002 was a dismal natural food year in Seattle that was associated with low crow productivity. To make matters worse, in some areas Red-tailed Hawks took most of the growing nestlings before they fledged. Pairs with helpers were noticeably more successful under these trying times. Helping was also most beneficial in the year with the lowest overall crow productivity in Michigan. Last, allowing some young to remain on the home territory may be a form of extended parental care. Those young least likely to obtain a territory on their own may be allowed to remain at home, where their survival is likely to be high. Like money in the bank, they mature socially to the point where they repay their parents by reproducing later when their chances of gaining a distant territory are greater or after they have enlarged the natal territory to the size where a portion can be budded off.[12]

Cooperative breeding appears to be even more complex and intriguing in Carrion Crows. Vittorio Baglione and his colleagues studied this European crow in Spain and found that some helpers simply remain on their birth territories to help their parents, as we observed for American Crows. Others disperse widely to find other relatives to help. Young male Carrion Crows may travel over thirteen territories to find a relative to assist. During these travels, they prospect in many territories but eventually choose, or are allowed, to settle with relatives, often their older siblings. The indirect genetic benefits of aiding kin appear to favor this helping because male helpers enhance their relatives' reproduction and even share mating with resident female breeders.[13]

Variety is truly the spice of crow life. The social scene for American Crows in Los Angeles, for example, appears to be unique. Most strikingly, the Hollywood crows build their nests close together in loose colonies, tolerate their neighbors' trespassing, and favor female-biased helping. Colonial nesting has been reported from several western locales and in populations of Northwestern Crows, but the females rarely remain at home and help. In birds, especially cooperative breeders, young females disperse earlier in life and to more distant locations than males. Young males that delay dispersal usually help. Female California crows act more like typical mammals; they delay dispersal and are more likely to help their parents than are males. Young males may remain on their natal territories and not help or

Red-tailed Hawk about to eat a crow nestling

they may disperse. Dispersing California crows are inclined to form non-breeding flocks that wander, but not too far from their home. It is tempting to say that young California crows are lured together by a force analogous to the surfing culture that lures other young Californians, or perhaps flocking is a response to rich resources. California crows that were studied lived on a food-rich golf course. Really, though, we just don't know.[14]

Mated adult crows interact regularly throughout the year by calling softly to each other and giving each other a good deal of physical attention by gently grabbing bills and grooming each other's feathers. Corvids groom each other to an unprecedented extent among birds. This "allopreening" is common in primates but not in birds. In allopreening, one partner edges up to the other, bows its head slightly, and raises its head feathers. Using a dexterous beak, the partner then grooms its mate by picking through the feathers just like a deftly fingered primate grooms its social companions. Allopreening is usually confined to the top of the head and facial area and serves the utilitarian function of removing parasites from hard-to-reach places. Allowing another bird so close when the recipient is vulnerable, however, suggests that this behavior may also provide an important social function related to maintaining the pair-bond. Allopreening can occur between birds perched on wires, in trees, or on the ground. If on the ground, rudimentary "tools" may also be used. A female American Crow in Florida was observed standing on dried cow dung to get above her mate and preen him more effectively. Gently grabbing beaks, known as "allobilling," is less common in crows than it is in ravens. It involves mutual mouthing that often escalates into sharp jabs and brief fighting.[15]

Even established pairs of crows court each spring. Males display their flying skills with acrobatic dives and rolls above observant females. Females get much of their food from males at this time. Mates and even some helpers feed begging mothers before, during, and after egg laying. This may be confused for allobilling, but it actually involves transfer of food. Courtship feeding is beneficial for both partners because only the female can incubate, so by feeding her the male ensures constant heating of the eggs, a healthy and surviving partner, and possibly more eggs. A well-fed female can lay a larger clutch. All associative, intimate behaviors like these become increasingly frequent during the breeding season, when maintenance of the pair is especially important.[16]

Crows, especially female crows, "like" foreplay, but what sex they have is quick. Ritualized displays precede sex in most birds, including crows. Lawrence Kilham observed over fifty American Crow precopulatory displays during the early 1980s in Florida. He describes the basic display as "the male, the female, or both crouched with body horizontal, wings out and drooping, and tail vibrating up and down." Females always performed the display, but males did not. Displaying crows also often vocalized, being especially prone to uttering soft *cu-koo* calls. Louder screams, audible from 275 yards (250 meters) were heard in nearly half of the copulations. In three instances males or females held sticks or other objects in their bills during the display.[17]

Copulation usually occurs on the nest but can also occur on the ground and, less frequently, in the branches. Male crows, as in most birds, do not have a penis, but their cloaca, the common outlet for excretory matter as well as sperm, swells during the mating season to facilitate fertilization of the female. Copulating crows simply touch their cloacas together to transfer sperm from the male to the female. To do this the male must get his tail under the female's. Copulations occur throughout the egg-laying period, but

Paired male and female American
Crows in copulatory sequence:
(a) precopulatory display; (b)
mounting; (c) insemination

each event usually lasts less than fifteen seconds, just long enough to ensure that each egg is fertilized. Kilham describes the full display and copulation sequence in a pair of American Crows. He observed: "A female was giving slow *caw-caw-caws* on her nest on 13 January when her mate came to the rim and mounted. I heard a single *cu-koo*. The male settled, waving his outspread wings which came to hang over the edge of the nest. The female stood up beneath him, her tail vibrating up and down as he worked his tail under hers. On the next day, the second of egg-laying, he again came to the rim, placing a foot on her neck before mounting. This time her body sank low as her head tilted way back. The crows vocalized (bills open) so loudly that they were audible at 250 meters." It is not always the male who mounts the female. Females mounting males have been observed, albeit only very rarely, in American and Northwestern Crows.[18]

Copulations do not only occur among paired males and females. Matings called "extrapair copulations" are sometimes seen. Kilham observed male American Crow helpers copulate with female breeders on two occasions. In one case, a "sneaky" helper responded to a female in precopulation display near the nest. The breeding male, who was nearby and the likely intended recipient of the display, immediately knocked the sneaker off the female and copulated with her three times over the next twelve minutes. This unusual rate of mating was probably the male's attempt to increase his sperm's chance of outcompeting any that the helper might have transferred during the brief extrapair copulation. In the second case, the helper was anything but sneaky. He joined a copulating pair at the nest, mounting the female at the same time the breeding male did. Genetic analysis has not been done on crows to determine the rate at which young are fathered by males other than the "breeding" male, but if investigations in other birds are representative, helpers and possibly floating or neighboring males occasionally father young with another male's mate. We suspect this is rare in crows, because the male remains close to his mate during the fertile period. The potential to reproduce certainly cannot be ruled out, as one of the reasons young males often remain as "helpers." In fact, in some species like those Spanish Carrion Crows, shared breeding may be a primary benefit of helping.[19]

Crows and ravens are famous for getting into weird situations—we've seen them enter homes, ride ferries, and invade grocery stores and fish-processing plants—but some of their natural behaviors are even more

When crows are sunning, they appear blissful

perplexing. On warm, sunny days crows will orient themselves perpendicular to the sun, spread their feathers, droop their wings, and lie prostrate on the ground in what looks like a semiconscious state. This odd behavior is common during the late summer molting period or after a prolonged cold, wet spell. Their partially closed eyes glaze over, and their mouths may open to help cool their sun-drenched bodies. Whole family groups may litter the ground seemingly dead, but if you try to approach they quickly come to life and move away.

What are these sunbathing beauties doing? Apparently, the solar radiation turns the birds' natural preen oil into a source of vitamin D. Ingesting it after sunning during normal preening activity could therefore provide an important dietary supplement. Sunning may also cause some skin and feather parasites to become more active and hence more accessible to preening birds.[20]

Crows sometimes sprawl out around the mounds of acid-producing ants. This craziness, or "anting," is a crow's version of an insecticide application. Anting crows grab ants with their beak, crush them, and wipe the natural oils across the underside of their wing feathers. Formic acid and pungent anal fluids squeezed from ants have insecticidal properties that may drive unwanted parasites from the bearer's body. During anting, live ants may also crawl on crows and actually eat some of the tiny troublesome creatures out of a beak's reach. But could anting have an even more subtle function? Some speculate that the formic acid squeezed from ants is intoxicating to birds. This and other signs of crow boredom led David Quammen to suggest that success has spoiled the crow. More precisely, he noted that crows "revel in formication!" Interestingly, in some corvids anting appears to be a learned behavior. As such, it could be an important component of crow culture passed down the generations of some family groups but not others.[21]

A company of crows applying crushed ants to their plumage

Crows play. Once thought to be reserved for people, play has been increasingly found in animals, especially long-lived social ones like crows. Young American Crows have been seen grabbing paper in the bill, leaping into the air, dropping the paper, and pouncing on it again. Our editor, Jean Thomson Black, vividly recalls a crow from her Connecticut backyard that would grab a child's foam toy and shake it playfully in its beak as if to beckon her to join in a game of tug-of-war. Flocks of crows and ravens often soar on windy days for hours on end in apparent play. They get lift from the wind's energy and ascend to perform loops, rolls, and dives. Ernest Good watched American Crows repeatedly fly upwind with considerable labor for thirty seconds or more then suddenly flip over, catch the wind, and hurtle back across the countryside. On the snowy hills of Maine, Common Ravens play with sticks, bones, and other objects. But what really strikes you is their preoccupation with "body surfing" on snow. We have seen them slide down a slope on their bellies only to get up, hop back to the top, and do it again. Lawrence Kilham observed several cases of apparent play among young American Crows in Florida. He reports crows hanging upside down from moss and swinging back and forth, routinely engaging in tug-of-war contests over inanimate objects like sticks and plants, occasionally taking sticks into the air where they are dropped and recaught before hitting the ground, and repeatedly rolling a dried raccoon skull down a stump. Kilham also saw playlike behavior between crows and other animals. A consistent theme in these interactions was dancing: crows apparently imitating the mating dance of cranes or jumping high over animals like calves and vultures. McCaw, Tony Angell's pet raven, routinely chased his husky dog. This sort of play, especially in young crows, may help sharpen coordination later needed to catch evasive prey, facilitate the development of social skills, teach them about their surroundings, or simply satisfy an inquisitive mind.[22]

As crows increasingly interact with people, they begin to play with some of our own toys. Carrion Crows and Jungle Crows steal and roll baseballs, tennis balls, and golf balls. Mostly this play is rudimentary and done by single birds. We had always suspected that such crows had mistaken these balls for eggs or nuts, but a detailed observation by Reiko Kurosawa, a Japanese crow researcher, suggests otherwise. Kurosawa was at the Tana City tennis court on November 25, 1993, in the company of about thirty Jungle Crows, when two of them approached one of the tennis nets. Eventually, one crow got on each side of the net and faced each other. A third crow flew in carry-

Crows at play often hang from branches

ing a tennis ball and threw it at the net. As the ball bounced off the net, the other crows watched it roll. Kurosawa also observed another crow bounce a rubber ball off a baseball net.[23]

What are these ball-playing crows up to? Are they just trying to crack open some new "nut" or "egg"? Or could they be mimicking human play? Kurasawa suggested the latter. If she is right, we might expect to see this play expand into a more interactive, social activity with crows passing balls among themselves. But if balls are just "tough nuts," then we would expect to see crows dropping them from greater heights onto hard surfaces or placing them in front of cars to facilitate cracking. That balls are thrown at soft nets suggests to us that these crows are really playing, not just foraging. The implications of this simple deduction are stunning; ball play is a primitive beginning of cultural transmission *across species*. The behavior of a few crows may have been diversified to include the ball-playing culture of people. If ball-playing spreads among crows by social learning, then we could conclude that crow culture had adopted an aspect of human culture.

Crow play involves tug-of-war

A mong the most conspicuous aspects of crow culture are their large and sometimes bizarre gatherings. We have mentioned that most crow species gather to roost communally at night. Between the staging beforehand, the roosting itself, and the noisy departure, roosts impress most observers and even occasionally drive nearby homeowners to desperation. The large winter foraging flocks of crows make grain farmers and Alfred Hitchcock movie buffs shudder. Noisy mobs that dive-bomb dangerous hawks and eagles cause us to pause and wonder. Long lines of flying crows are simply astounding. Consider one such line we saw flying parallel to our evening commuting route. The American Crows were flying at about twenty miles per hour, and our bus was going thirty miles per hour the opposite way. We watched this solid procession of crows pass our window for fifteen minutes, at a combined rate of fifty miles per hour, suggesting that the line of crows was at least twelve miles long and may have contained more than seven thousand birds. Scientific inquiry has helped explain most of these gatherings, but some require quite a lot of ruminating and exemplify the tantalizingly bizarre questions that lie at the heart of the crow mystique.[24]

Nothing raises people's curiosity about crows more than their dive-bombing of predators, including humans. Dive-bombing potentially dangerous predators at first blush seems foolish, but it has a variety of benefits. Hawks, owls, eagles, raccoons, cats, foxes, coyotes, and even people are often the subjects of mob aggression by hysterical crows. To crows these are all predators, having at one time or another been caught by a crow in the act of preying on nestlings or older family members. While calling loudly, members of the mob orchestrate strafing runs at the potential predator, often striking it. Blows so delivered can cut people, knock small predators from their perches, and may even kill some predators. We have found both Red-tailed Hawks and Sharp-shinned Hawks (*Accipiter striatus*) seriously in-

jured and grounded by mobbing American Crows. Crows can briefly para-
lyze squirrels by knocking them out of a tree. Even a Bald Eagle (*Haliaeetus leucocephalus*) was suspected of being killed by American Crows defending their nest. Despite these successes, mobbing can be deadly. Peregrine Falcons (*Falco peregrinus*) are known to quickly change their flight pattern to grab and kill a mobbing crow.[25]

Mobbing has many potential benefits that balance the costs of this risky behavior. Mobbing crows confuse and distract the predator, show it to their young, perhaps drive it from the area, or signal that the quarry is aware and informed and will not be surprised by a sneak attack. All of these benefits reduce the predator's hunting effectiveness and may increase crow survival and learning. The ability of mobs to move predators out of high-use areas is especially important for securing nightly roost sites. This is why owls detected during the day are vigorously mobbed—they are often forced to move so that when they begin hunting after the sun sets, they are likely to be far from vulnerable, sleeping crows.[26]

Mobbing may also confer more subtle benefits, like advertising the ability of a mobber to take a risk successfully. Such subtlety is suggested by the fact that crows often go out of their way to mob dangerous predators, especially when others are watching. Jays (*Garrulus glandarius*) have awakened to go directly to the territory of a Tawny Owl (*Strix aluco*) to engage the predator. On finding the owl absent, the jays flew directly to the next closest owl territory and harassed that sleepy resident. Could the mobbing jays be gaining status in the eyes of their flock mates? Tore Slagsvold, a Norwegian ornithologist, has presented convincing experimental evidence to suggest that mobbing Hooded Crows are in fact advertising their self worth! Slagsvold determined that the most vigorous mobbers were the largest, most dominant male crows. Dominant males may mob most vigorously because

A mob of crows dive-bombs an intruding Golden Eagle

their size or experience makes them least vulnerable. In birds, however, small size would actually be advantageous to a mobber since increased size reduces flight performance. Experience may still play a role, but the true intention of dominant male mobbers was revealed when Slagsvold presented a mounted Eagle Owl (*Bubo bubo*) surrounded by mounted crows to wild crows. Mobbers consistently attacked the *mounted crows* as well as the owl. In fact, they attacked the crows more than the owl when the experiment was conducted in common flocking areas, where advertising one's worth is likely to reap the greatest social benefit. In contrast, in breeding areas where moving a predator is most important, the crows attacked the owl more than the mounted crows. These experiments clearly show the subtlety with which crows can adjust their behavior to the situation at hand. It has been known for some time that crows mob predators, including people, most vigorously when danger is most apparent. A dead crow or even a black cloth near a predator stimulates extraordinary mobbing. But Hooded Crows have shown us that danger is not the only force motivating this risky behavior.[27]

When the clear and present danger of a predator is detected and alarm calls are voiced, the territorial boundaries separating breeding birds are dissolved and pairs of birds descend on the invader from all sections. One is reminded of how war tends to make allies out of enemies when a common benefit is to be realized. Hooded Crows also remind us that sometimes impressing one's "allies" is the true motivation for risky, cooperative behavior.

When crows mob people, our worlds literally collide. We rarely attract the scorn of parents tending a nest high in the tree canopy, but when young crows first fledge from the nest, they often spend a few days relatively flightless low in a tree, in a bush, or even on the ground. Even though passing humans may not notice these grounded crows, in the eyes of attentive parents the passers-by pose a real threat at a critical stage, and no holds are barred. This is when people get their hair pulled, scalps cut, and shoulders bruised by crows. Such defense of nestlings and fledglings is understandable and even admired by most people. But occasionally crows seem to just have it out for some people for no apparent reason. We often hear from red-haired women and balding men that they are swooped on by crows. A columnist for the *Seattle Times* claims that her neighborhood crows single her out for scolding. Perhaps the crows do not approve of her opinions, but it seems more likely that attacking crows have been harassed in the past by humans with similar physical, postural, or vocal features. Crows may remember such events and distinguish general features of people, so those of us unfortunate enough to fit the profile can be singled out for retribution. For this reason crow researchers often wear costumes when climbing crow nests to band young so that the parents do not fixate on our normal appearance and features. The hat-wearing coauthor of this book is quickly identified by resident crow families and, year after year, scolded for his attacks on the local crows who were harassing breeding Screech Owls (*Otus asio*) ten years earlier.[28]

More than two million American Crows have been seen roosting together in Oklahoma. Communal roosting, though spectacular, is easy to explain and not unique to crows. Biologists have studied the phenomenon for decades, concluding that roosts help crows keep warm, ward off predators, and pool information, particularly about finding food. Roosting is an obvious cultural attribute because young birds learn about roosts by following parents and because, as roosts change location, ignorant crows find them by following knowledgeable crows. More crows produce more heat and more effectively chase or confuse predators. Each individual crow benefits by joining a roost based on simple probability theory: if a predator will kill one crow each night, then a crow's chances of dying are one in a thousand if the roost has a thousand crows, but one in two if you roost only with your mate. Dilution of predation risk may explain why families in large communal roosts do not appear to roost together. It would be better to spread out just in case the predator takes more than one crow at a time or returns to hunt again at the place it was last successful. More eyes are effective in spotting predators and efficient in locating unpredictable foods.[29]

Roosts have been christened "information centers" because they appear to serve as debriefing arenas for hunters each evening. Birds were thought to disperse in many directions from the roost each morning to hunt and forage rather independently. On arriving at the roost in the evening or as they left the next morning, successful hunters would divulge their secrets vocally or with a variety of postural displays and lead unsuccessful hunters to food. Common Ravens do exactly that. Young, vagrant, nonbreeding ravens that find large animal carcasses are rarely allowed to feed on them alone because of defensive resident territorial adults. So these vagrants return to communal roosts and inform others about newly found distant foods. It took Bernd Heinrich and Colleen and John Marzluff years sitting atop trees and huddling in small natural blinds exposed to the winter cold of New England to unravel the mysteries of raven roosting. We captured ravens, radio-tagged them, and released them at dead moose, frozen cows, and piles of slaughterhouse scraps. We followed the ravens and their newfound knowledge to roosts where they initiated social soaring displays to move their roost mates near the carcass that evening or simply flew purposefully the next morning in the direction of the carcass. We ruled out the possibility that naive birds simply smelled food on our birds' breath. Yes, carcass breath is strong, but our released birds often did not eat before they went to roost. Because only

A gang of ravens on a roadkill

our birds knew where we placed food in the snowy hills and we could track them from food to roost and back, we proved that knowledgeable birds led naive ones to new foods.[30]

Detailed observations of ravens from Wales confirmed and extended our findings. Dominant ravens often initiate social soaring displays after they find new feeding locations and just before they lead their roost mates

Crows roost for the evening

to the new food. As a result, all the ravens in roosts, from fifty to five hundred of them, arrive nearly simultaneously at the food bonanza, overpower the territorial adults, and eat their fill. This works for ravens because all roosting individuals benefit and because foods are rich but hard to find. Leaders gain access to previously defended foods and followers find new food. Unpredictability favors information sharing, as Geir Sonerud and his students have clearly shown in the snowy forests of southern Norway. There, Hooded Crows are most likely to join communal roosts after heavy snows and thereby learn about new feeding opportunities. Snow literally erases many feeding locations and forces the crows to rely on roost mates for information.[31]

Ravens and Hooded Crows need information to track unpredictable foods, and ravens need a small army, at least nine birds, to access it. Other crows, however, may need neither. Dumps, dumpsters, and agricultural fields provide consistent feeding grounds that are usually too large, or located in inappropriate areas, to be defended. As a result, detailed studies of radio-tagged American Crows in New Jersey by Don Caccamise and his students failed to demonstrate that crow roosts function as centers of foraging information. Instead, they appear to be efficient gatherings of individuals that have been foraging on nearby rich foods. In New Jersey, this was New York City's refuse dump on Staten Island. Crows foraging at the dump, even for a day, saved commuting time and energy by sleeping near their food.[32]

All crows in a roost may not be there for the same reason. A recently widowed crow may join a roost to find a new mate; a watchful youngster may learn subtle social skills from surrounding adults; a recently arriving migrant may join to find food; or a wise territorial bird may join after a large snowfall for warmth and extra safety. European Hooded Crows reap such multiple benefits, and we suspect other crows and ravens do, too.[33]

The benefits of roosting in large groups, especially in cities, bring bands of noisy crows into direct conflict with people. This is providing some new and dramatic costs to communal roosting and changing some aspects of crow culture. People often tire of nearby crow roosts and undertake herculean efforts to disperse them. In the most blatant attack, 328,000 crows were killed in 1940 when a Rockford, Illinois, roost was blown up with dynamite. Smaller, less conspicuous reductions occur today, indicating that crows push their limits with people when they amass on our turf. Crow culture changes in response. Ernest Good, one of the first biologists to study crows in detail, noted that in the 1950s in Ohio where crows were often hunted at their rural roosts, birds would gather in later afternoon at a few places several miles from the actual roost. They then flew in small groups to the roost in the last hour or so of daylight. Others, ourselves included, observe similar pre-roost gatherings in and around urban areas. Gatherings before roosting may also help birds get an easy meal before the long night, but if this were the only reason for gathering, we would expect crows to actually roost at these afternoon snacking sites. The fact that they gradually leave (maybe even *sneak* off) to converge at a distant site suggests other motives, such as predator avoidance or information exchange.[34]

Even small assaults on roosting crows can change local culture. Near the Snohomish River in western Washington, longtime farmer Bob Ricci told us that he shot three crows one evening in 2002 as they were marauding his sprouting corn. He had to lull them into a false sense of security by shooting discretely from his tractor rather than obviously from his truck or on foot. But this discretion did not let his actions go unnoticed. From that day on, the crows never bothered his corn again and the evening flight lanes to the roost were altered to go around, rather than over, his homestead, as they had done for years. Avoiding Ricci's corn represents social learning because all of the thousands of crows that now follow this path to the roost could not have directly experienced the farmer's wrath. The new travel route was simply adopted by naive birds following experienced birds. Most crows avoided Ricci's corn for a year, but a few visited in 2003 and were shot. Changes in tradition may require several trial-and-error and social learning reinforcements before they actually become engrained cultural differences.

Common Ravens in southwestern Idaho may have learned enough to evolve a new roosting culture. These ravens act like Staten Island crows, feeding on predictable agricultural products like corn, feedlot waste, small rodents, and bugs. They roost in huge numbers and do not regularly recruit roost mates to new food bonanzas. Recruitment is not necessary for open-country ravens. In the open sagebrush habitat that they inhabit, few secrets are possible. A hungry raven needs only to perch on an outcropping, water tower, or power line to survey the landscape and spot others at new foods. Perhaps the culture of sharing information at communal roosts has waned, in part because the culture of people over the past fifty years has provided plenty of food in reliable locations. Together with an environment that allows birds to easily find new foods, ravens can forage for themselves. Roosting culture and agriculture have coevolved.[35]

On a large tarmac parking lot next to the University of Washington's football stadium a most bizarre gathering of American Crows occurs each morning. Before daylight, birds begin leaving their communal roost two miles to the south. Guided by the lights of the parking lot, they arrive in loose flocks and strings of ten to one hundred birds to gather in the trees and asphalt of the lot. The cacophony of calls is deafening, making it hard to concentrate on the growing black tide. Wave upon wave of birds continues to fill a large section of the parking lot. In the dim light, you can wade through crows indifferent to your presence. They walk away as you approach but appear to be in no hurry. Their business is here on the ground. Yet there is no food, no permanent water, nothing but black macadam.

We've spent many mornings attending this gathering but have little rational explanation for it. Our students have measured temperatures on and around the lot, but it does not appear to serve as a heating pad ready to warm cold crow toes. Maybe crows just attempt to avoid drawing attention to the roost site in the morning, as it also seems they do by gathering before they roost each evening. We speculate, however, that crows are simply gathering to greet each other, to get organized back into flocks or family groups split apart for safety's sake in the roost, or maybe to get their juices flowing by stretching, yelling, and screaming. Could they be crow pep rallies? Does the social stimulation of the gathering charge the system like caffeine at a local coffee shop? They regroup, catch up on the latest buzz, prepare for the day's events, and shake the sleep out of their bones. Yet there is even more. When describing this scene to a long-time Seattle resident in 2002, he quipped that the parking lot we were at used to be a dump where crows gathered to plunder garbage each morning for decades. That would explain the location of the roost: close to a rich food supply. But it also requires amazing tradition and culture. Crows still roost there, even though the food is gone, because they always have. It is tradition. The location is not especially useful as a commuting stopover anymore because the dump is gone. Yet this ritual has persisted for over forty years, passed on to at least four generations of crows, just for old times sake, so to speak. Even more amazing, then, is the daily homage paid to a former feasting site. This gathering may have had an important foraging function in the past, but now it appears highly ritualized. This could be why it is difficult to explain on the grounds of present value; new values like group organization, awakening, and social greeting are just beginning to emerge, but past values continue to shape these crows' behavior.

Traditional gatherings at places like old dumps are mundane relative to stories about crows gathering to aid the injured, bid fallen comrades goodbye, or to execute a group member. Yet careful observers claim to have seen such events. When we visited a roost that included some sick and dying

Overleaf: Crows at muster. Thousands gather before dawn at an old dumpsite, now a University of Washington parking lot.

American Crows, we noticed that apparently healthy crows were often vigilant near the sick. Kevin McGowan has seen similar behavior and noted
that healthy American Crows would feed sick family members. Even handi-
capped associates are cared for. Scientists in British Columbia recorded in
detail how a crippled and partially blind Northwestern Crow was fed, and
hence kept alive, by its group members.[36]

"Funerals" and "executions" are even more bizarre. Both sorts of observations begin in a like manner. Someone realizes that a lot of noise is being made by hundreds or thousands of crows at a place or time where there
are usually just a few. The noise and gathering continues for some fifteen or
more minutes, then there is silence, and the crows are gone. On inspection,
a dead crow is found at the gathering site. The noise is unusual for its timing and location, but it is the *silent departure* that seems so uncrowlike to
witnesses.

Beth Wapelhorst, a resident of Seattle, has described several encounters with dead American Crows. All were accompanied by great commotion, then silence. In contrast, we have seen several predation events where
hawks or people kill crows. These have lots of commotion as crows dive-
bomb the assailant, but they never end in silence. Rather, they end with the
noisy mob's chasing the departing hawk or uttering specific "dispersal calls"
(see chapter 6) that signal the group to depart. So the silence after gathering
strikes us as different. Perhaps it is a gathering for a dead group member, not
the culprit responsible for the death. Even so, such gatherings need not invoke supernatural explanations. Crows may simply gather at a dead crow to
make associations between a place or a setting and danger. Group members
might be learning to avoid an area or others like it in the future for their own
safety. Such learning could explain why crows stop frequenting areas where
they have been trapped or killed. The resident pair of crows at Marzluff's,

A vigilant crow watches over a dying adult

for example, did not enter the yard for nearly six months after they were trapped and banded, despite daily visits before the capture.

To satisfy our curiosity about the possibility of crow funerals, we placed dead American Crows in our study area. The response, though predictable, was a bit disappointing. Resident adults first saw the dead birds—in a matter of minutes—and gave assembly calls that brought all those in the area to the scene in a beeline. Ten to twenty birds were soon circling low over the carcass, scolding loudly. A few landed near the dead bird and scolded. Those approaching on the ground were the territory owners. Perhaps they were making sure that the dead bird was not one of their fledglings or a sleeping intruder. Maybe they were showing off to the others. The response was the same as we see when we trap a resident under a net to band it. The trapped crows appear dead under the net, and a swarm of scolding crows quickly assembles, then disperses. In our experiments crowds dispersed after about thirty minutes. We even were able to place a pair's recently dead fledgling back in its territory and the result was the same. There were no funeral processions. No sudden moment of silence for the fallen. Just a straightforward response by vigilant animals to a scene of possible danger.

An observation harder to explain away was made by Lee Bond, a retired oceanographic technician from Seattle. Bond was attracted to the din of about a thousand American Crows crowded shoulder to shoulder on power lines and trees at his neighborhood park one autumn afternoon. It was too early for a pre-roost gathering, and Bond had never seen such a mass of crows in this area. Suddenly a movement on a wire about twenty yards (meters) away caught his attention. As he looked at the row of crows, one fluttered dead to the ground, its neck apparently broken. The entire murder of crows left silently. To Bond it was a memorable, important, and eerie sight—the silent departure of so many crows after one fell dead. Other than the displaced head suggesting a broken neck, there were no obvious wounds, burns, or abnormalities on the crow. Bond is certain that the movement he saw was one crow killing another. Apparently this was an execution in the presence of a large group that was there before, rather than in response to, the death. They were witnesses rather than respondents.

We are hard pressed to dismiss this observation but equally hard pressed to explain it. We could imagine a crowd of crows foraging on autumn's bounty of acorns suddenly flying onto branches and wires in response to an alarm call. Perhaps a predator dashed through, wounding a flock member who gripped the wire momentarily. Or perhaps in the confusion, one crow contacted a ground and live wire and got a killing shock. Then as Bond passed, a wounded, recently killed, or just naturally expiring crow fluttered to the ground. These explanations are plausible, but not very likely. Several things don't add up. The departure of a predator would have elicited a noisy pursuit. Electrocution would probably have killed more than one crow, and burns would have been evident. Prior death would likely have resulted in a corpse on the ground, not on the wire. A dead crow on the wire would be stiff and stuck to the wire, not limply falling to the ground. In addition, the role of human intervention is unlikely; no people were around to be accused of fowl play. So, was this a crow execution? Executions are unknown outside of our species. What could a crow do to warrant a death sentence? Perhaps it was an accidental result of territorial combat gotten out of hand or an over zealous fight for increased social status?

An observation Marzluff made as a graduate student nearly twenty years ago on a pine-studded mesa in Flagstaff, Arizona, sheds some light. He interrupted a killing in progress. About thirty American Crows were calling loudly as one grounded crow was repeatedly attacked by others. The grounded crow had an obviously broken wing and could not fly. He suspected that it had been hit by a car and was being taken advantage of by a rival. Rather than wait for the inevitable death, Marzluff caught and euthanized the crow so that it could be mounted and used for some planned experiments. Kurt Kotrschal, director of the Konrad Lorenz Research Station in Grünau, Austria, suggests that cars need not be involved. He and his colleagues observed Common Ravens socially mobbing a raven whose wing was broken, not by a car, but by the violent raven mob that had gathered. But these events and those relayed by Lee Bond suggest that crows can occasionally kill each other. They also allow us to reject several plausible ideas for the killing. Killing does not appear to be motivated by hunger, because the dead are not eaten. It is not done in the course of usual territorial defense, because too many bystanders are involved. Perhaps it is done for status enhancement. A subordinate that kills or partakes in the killing of a dominant may rise in status among its peers and is therefore climbing the avian social ladder. A sort of "gang mentality" may actually drive this extreme behavior. As many gather, several move in to kill an injured or newly vulnerable dominant bird. The competitive crowd increases motivation to kill. As one crow gets close, another follows or moves closer, causing the first attacker to get still closer or peck still harder. This grows like a rolling snowball until one bird kills another. This would have happened in Arizona had Marzluff not intervened. It happened in Austria. Was it possible in the scene Bond described? If the birds had been fighting for some time before his discovery, the loser's distress calls would have assembled the apparent witnesses, leaving Bond to see merely the final blow in a long chase. Possible, but it's a stretch—why would so many crows be interested in a fight between upstarts or a prolonged vying for territory? Maybe this was an extreme form of crow justice. In any event, it seems most plausible for crows to suddenly and silently leave a dead bird only if it was killed out of their sight or actually killed by them. More observations from interested people like you are needed to resolve these intriguing issues.[37]

We are hopeful that the gathering behaviors of crows can be used to point out the inconsistency with which people often react to wildlife. Many stories about funerals or executions are immediately dismissed as impossible for animals other than humans. But the frequency of such stories and the astonishment of the witnesses suggest that social species like crows may employ some social practices not entirely unlike our own. Roosting crows also demand us to dig deep into our souls and reconsider our ethics toward other forms of life. Roosting flocks elicit conflicting emotions: anger, wonder, concern, admiration, and curiosity. The closer we live to large aggregations, the more likely we are to be angry and concerned, but from a distance thousands of synchronously swirling crows are one of nature's marvelous mysteries for most of us. As a society, we must come to grips with these challenging and conflicting views of other species. Appreciation, wonder, astonishment, and careful insight would seem prudent.

SIX

Communication and Culture

The afternoon calm is disturbed by a growing growl that swirls above the snowy conifer forest. As the sound gathers force, black forms wing in from all directions. Judging from the near-hysterical pitch of their calls, crows are mobbing something big and dangerous. The thunderous wall of sound moves in our direction as an eagle flies overhead followed by a frantic black mob.

We are often alerted to spectacular sights like marauding eagles by listening to the voices of crows and other animals. Crows communicate their motivations, identities, and report on local conditions each time they *caw* or *croak*. But sounds made by crows are much more than simple signals that we occasionally intercept. Together they encode a complex language, steeped in cultural tradition. Cultural evolution likely produces calls unique to the

Crows in hot pursuit of a fleeing eagle

social group, suggesting specific predators and local activity. Listening to a crow provides insights into the crow's ability to recognize partners, relatives, and enemies. Crow vocalizations indicate the variety of ways these birds interact with their world and show us how perceptive they are in understanding their social and physical environments.

Yet is it legitimate to think of a crow's *caw* as part of a "language"? *The American Heritage Dictionary* defines *language* as "any system of signs, symbols, etc., used for communication." Scholars of animal linguistics demand more. To be language, animal communication should consist of symbolic referents, the rules of syntax, and the ability to create new symbols and use them in novel and appropriate situations. Few animals come close to the human language's symbolism, syntax, and creativity, but we suggest that most corvid vocal communication systems meet at least the basic standards of a language. The clucking of Common Ravens symbolically represents a dangerous invader near the nest. But their trilling indicates a trespassing challenger. American Crows represent distant predators with *ko* calls. These vocal symbols have clear and consistent meanings, but they are arbitrary (that is, symbolic): Why has a trill come to represent a challenger rather than a predator? Not only are calls symbolic, but as we learn more about corvid vocalizations we are recognizing basic rules of syntax. The American Crow's *caw*, for example, is doubled in rapid succession to emphasize territorial defense, and Pinyon Jay alarm calls are combined with wavering contact calls to indicate increasing excitement by jays mobbing predators. The importance of syntax to the meaning of American Crow kaws has even been demonstrated experimentally. Psychologist Nicholas Thompson of Clark University recorded caws from wild crows, then remixed and replayed them to crows. Depending on how he arranged a sequence of cawing—as a structured versus an unstructured composition—Thompson could attract or repel crows *with the same caws*. Obviously, syntax, not just the symbol, makes a difference to the meaning. Still, many linguists would argue that these rules of grammar are too simple to qualify crow communication as a language. Human grammar regularly allows words to be rearranged and change a sentence's meaning. But human grammar also allows us to anticipate connections among words separated by other words of vastly different meaning. Words like "if" and "then" are an example. We do not know of such "recursive" properties in animal communication systems. Most corvids quickly invent new vocal symbols and use them in appropriate and novel situations. One way they do this is by mimicking the sounds of other animals, inanimate objects, and human speech.[1]

Crow and raven vocal communication possesses the basic aspects linguists require of a language. But let's temper our interpretation a bit. Individual caws, *quorks,* and screams may often express general emotion or mood, like the grunts, cries, and screams of humans. Some calls or phrases have discrete meaning, like human words or sentences. They may be combined in complex ways and given in a variety of contexts to vary and enrich their meaning, but these are probably not constructed with the socially approved rules of grammar that we use to compose sentences and paragraphs. Rather, crows say what they see, feel, or need in a way that conveys information for their own benefit. Natural selection is their editor, and a ruthless one at that: extra noise could attract unwanted attention from a predator or persecutor; the wrong expression could provoke combat. Crows can't afford to be eloquent, obtuse, verbose, or flowery—nature won't tolerate it. Only when the individual is protected within the mob or in a very comfortable setting does a crow really yammer.

Not only do crows and ravens possess basic language skills, they also have large vocabularies. In Europe, the Common Raven utters nearly eighty distinct calls. An individual raven does not use all eighty. Individual birds utter about twenty basic calls, but each is variable in intensity, duration, pitch, and rate. Many vocalizations are individually distinct so that the caller's aim and identity are stamped onto each utterance. In America, we know that male and female ravens share some but not all their calls. Paired birds coordinate calling bouts to emphasize their territory and keep track of each other across large home ranges. Territorial breeders and vagrant nonbreeders also have distinct voices used to rebuff intruders and attract allies. Simply by listening to ravens, one becomes aware of gender- and age-specific roles and the importance of the bond between mated birds. We can also infer that crows report on distinct aspects of their environment—their mates, their intruders, predators, and foods. They say who they are and how they fit in.[2]

Raven vocalizations tell us about the bird's complex culture. Raven mates share some calls not used by others, and neighboring males and females share many calls, some of which are not used by more distant ravens. Mated pairs develop unique vocal repertoires by copying a few of each other's calls. Ravens within a geographic region come to possess unique dialects as males copy some of their neighboring males' calls and females copy a few of their neighboring females' calls. This sharing of calls by neighbors and mates can produce sex- and pair-specific calls, as well as geographic dialects. The fact that such distinctions occur and are produced by social learning illustrates how cultural evolution can shape an essential and defining part of the raven's character.[3]

Ravens have shown us that decoding another animal's communication system is a powerful way to know another species. We can't talk with most animals, but in crows, we can start to eavesdrop. American Crows demonstrate the complexity, variety, syntax, and individuality common to all corvid communication systems. They clarify their rich array of vocalizations with postures, gestures, and facial expressions to communicate their perceptions to a wide audience.

C rows utter a variety of sounds that vary in duration, pitch, harmonic structure, and vibrato. By recording vocalizing crows under a variety of situations, analyzing these calls with a computer, and playing different calls to wild and captive crows, researchers are beginning to understand these call elements. An exact count of the numbers of types of such calls is difficult because crows use diverse dialects across the country and the various researchers use different descriptions to catalog crow calls. One study documented twenty-three types of calls, many of which Cynthia Sims Parr later subdivided in her splendid investigation of American Crow cawing. Taken together, these studies suggest that American Crows have upward of thirty or more distinct building blocks for their communication system. Crows use these basic elements singly or combine them into series.[4]

Nearly every crow linguist recognizes eight functional groups of calls. American Crows of both sexes appear to give all call types. Some calls are hard to categorize because their pitch and length vary with the calling crow's changing emotion. Despite this, all American Crow caws have clear differences.

Crows give at least three distinguishable *assembly caws* that are associated with mobbing, scolding, and diving at potential predators. These "two-syllable caws," "long caws," and "harsh caws" focus American Crows on the object of the caller's scorn—usually a stationary cat or owl—in an attempt to drive the creature from the area, alert others to its presence, or perhaps teach family and flock members about dangerous situations. A variety of playback experiments confirmed what crow hunters have long known; if you broadcast these

Calling crows

calls to crows, they will come flying toward you nine times out of ten. Even French corvids understand the American Crows' message.

If assembly caws are the "on" switch for American Crow attacks, then *dispersal caws* are the "off" switch. Harmonically complex, initially inflected, *koaws* are the most stereotypical dispersal call. These are given in response to immediate danger and cause birds to fly away from the scene, although they may initially attract crows to see the predator before they disperse. Researchers in Pennsylvania, Maine, and Virginia have successfully told American Crows to scram by playing this call.[5]

Koaws are often intermixed with or blended into harsher assembly caws. They are given during territorial defense in addition to mobbing. These facts led Cynthia Sims Parr to a striking conclusion: koaws may be given to intruding crows, not predators. American Crows leave at the utterance of a koaw not because of a dangerous predator but because of threat from a peeved territory owner. This makes sense in light of an early observation that the safest way to leave a dangerous predator is simply to fly *silently* away. Others can then follow without attracting undue attention.

The *ko,* or *warning caw,* alerts crows to impending danger. It is a sharp, tonal caw given to predators heard or seen from a distance. The ko alerts other American Crows to the presence and general location of a predator. Kos may combine with harsher and longer assembly caws as the predator is finally seen or approached more closely. It is possible that the seriousness of the threat or even the type of predator is encoded in the intensity with which kos are delivered. Low-flying hawks in particular elicit these calls. Field researchers can often guess with good accuracy what crows are mobbing based on the intensity of their caws, suggesting that they may be identifying the intruder as well as scolding it. Crows do not appear as adept at naming predators as are some monkeys. Vervets give alarm calls that refer to the type of predator so that listeners can run up trees to avoid snakes and run down trees to avoid eagles. Crows may not need such specificity because they approach and mob most predators.[6]

The largest missing piece of the crow's Rosetta stone is the chunk that tells us how the familiar caw is used to alert, warn, contact, and greet others. As we have seen, recognizable variants of caws (kos, koaws, harsh caws) have fairly consistent functions. But generic caws of long (more than 400 milliseconds) to short (150 milliseconds) duration are given in a whole host of settings, often accompanied by postural displays, to evoke a variety of responses. Caws of all durations are used to defend the territory, but their coupling with descending glides as the wings are held up in a U-shaped, or dihedral, position is typically used to announce one's arrival at a gathering. American Crows cawing their ownership of territory flick their tails by quickly and repeatedly spreading and constricting their tail feathers and pump their upper bodies up and down while slightly flexing their closed wings. As they pump up they caw as if to push the noise further or add a visual exclamation point to their territorial pronouncement. We suspect that, like human voices, all caws

Dihedral wing display used by crows arriving at a gathering

are individually identifiable. Such means of individual recognition is common in other flocking corvids and helps coordinate important social activities like group defense, mobbing, helping, group foraging, and communal roosting.[7]

Don't try this at home, but if you hold an adult American Crow upside-down by its feet and shake it, it will give a unique squall or scream that quickly attracts others to the scene. Squalls grade into harsh death cries if a struggling crow is dying. Hunters use variants of these distress calls to attract more victims, and wildlife biologists use them to disperse roosting crows from unwanted areas. You can purchase commercial recordings of these calls to keep crows from sensitive areas in your yard or just play mind games with your feathered friends. Broadcasting distress calls immediately attracts crows to the source of danger but then causes them to quickly, and often silently, disperse. They will usually not return for weeks, if at all, apparently because they have associated the location with a serious crime. In this way, crow language facilitates cultural evolution; calling crows attract naive birds to a dangerous location where they learn *socially* to avoid a place, person, trap, or predator. Such social learning likely accompanied the "execution" we described earlier, which would produce the silent departure and lack of return to the area after the crow was killed.[8]

Juveniles and incubating females of all crow and raven species noisily utter wavering calls while leaning forward with mouths agape and wings flapping. The sight or sound of a parent or mate approaching with food triggers this response. This begging may increase in intensity to indicate the degree of hunger or the proximity of the impending meal. They also likely alert parents to the movements of wandering juveniles and may identify individuals so that food is not wasted on unrelated birds when family groups intermingle, as they often do in late summer or autumn. Garbled begging

Pumping postures emphasize cawing

calls can often be heard as a youngster has a plug of food thrust into its open mouth.[9]

Short-distance vocalizations often are given by crows who bow in a stereotyped way to each other. Bowing crows shake their heads, pull their heads to their breasts, and lean forward. They might groan, snap their bills, or open their bills in apparent silence during these displays. We have seen ravens do similar displays that may signal an impending attack if given to a rival or precede sidling toward and allopreening if given to a social partner.

Crows and ravens make a variety of soft squawks, coos, bill clicks or snaps, rattles, cackles, and growls as they sit alone or close to a mate or family member. The unstructured "gargling" or "rambling" soliloquies of young, unmated American Crows and Common Ravens involve combina-

tions of these and mimicked noises from their locale. These elements may be combined during courtship or crafted into group-specific signatures by adult crows. Courting crows usually sing from the branches of trees. Crows occasionally sing while flying. The naturalist C. W. Townsend described crow courtship in the early 1920s: facing the female, "the male bowed low, slightly spreading his wings and tail and puffing out his body feathers. After two bows, he sang his rattling song, beginning with his head up and finishing it with his head lower than his feet. The whole performance was repeated several times. The song, such as it was, issued forth during the lowering of the head." Townsend later noted that *coi-ou* or *cou* calls often followed the rattle. On one occasion he counted fifty-four rattles "in succession followed by a series of *cous*." The bowing, rattling, and cooing of crow courtship is a subtle but common part of adult crow chatter during early spring.[10]

A more general and intriguing function of American Crow "song" was found by Eleanor Brown, a behavioral scientist who carefully recorded crow sounds in a variety of experimental settings. Captive crows sang complex mixes of coos, rattles, soft caws, growls, and unique noises that were consistent within, but different between, social groups. Crow groups learned their songs through mimicry of unique elements and phrases by newcomers. Mimicry is a form of social learning that allows song to evolve cultural dialects. Group members use dialects to recognize one another. Moreover, because new members adapted their songs to that of the group, the group's song could persist for generations. Just as ravens culturally evolved geographic dialects, American Crows may culturally evolve local dialects. Many

Nestlings beg from an approaching parent

details about the cultural evolution of vocal repertoires in crows and ravens require further research, but our current observations show that, like people and other primates, crows and ravens culturally create and keep dialects.[11]

Crows and ravens enhance their natural repertoires by mimicking the sounds of other animals, including people, and inanimate objects. Don't be surprised to discover that the dog you heard barking, the owl hooting, the child screaming, or the chicken clucking and crowing was actually a crow. Crows have been recorded mimicking all of the above. Crows can also speak fine English. Their complex throat muscles allow them to verbalize such phrases as "I'm Jim Crow" or "Oh my God, oh Lord." This they do expertly without the need for their tongue to be split, a cruel and unfounded suggestion made two thousand years ago by Pliny the Elder. As Pete Byers writes in his book *The Lost Folk Art of Crow Taming,* all you need to get your crow to talk is "patience, perseverance, and repetition." His Ohio-born crow, Edgar, imitated the cheering, hooting, and clapping that accompanied a television game show. Mickey, the crow at the National Aviary in Pittsburgh, greets visitors with "Hey bro, what's happenin'?"[12]

Mimicking need not be mindless copying. Konrad Lorenz's pet raven Roah learned to mimic his name and then used it in new and appropriate contexts. When Roah perceived that Lorenz was heading toward a dangerous situation, he would fly in another direction and repeatedly call "roah, roah" to lure Lorenz to safety. A neighbor's Hooded Crow, "Hansl," returned one day with a broken toe and spoke in German rather than using crow calls. What he uttered was equally impressive. Translated, he said, "Got 'im in t' bloomin trap!" He may have mimicked what the trapper said. But clearly, Hansl the crow made it known to Lorenz's neighbor what had happened to his toe. These are amazing examples of the creation of vocal signals used in new and appropriate situations. Surely crows and ravens do this with sounds other than those mimicked from people, such as those of important social partners. But when they mimic us we can easily appraise their true language skills. Use of mimicked sounds by birds is analogous to the use of sign language by apes; both allow us to better understand the comprehension, creation, and use of symbolic communication systems by nonhumans. Mimicry, in general, could be an important way for crows to expand their vocabularies and develop unique local dialects.[13]

Two aspects of American Crow caws show interesting parallels with other species. First kos are the most tonal, and hence the most difficult, call to locate. A crow giving a high-pitched, uniform ko is much harder to pin down than one harshly scolding a predator. Appropriately, kos are used to indicate the presence of a distant or unseen predator, often a hawk, without giving the caller away. Second, unstructured mobbing using a combination of call types includes harsh calls that are easy to locate and help the mob gather. Because they emphasize lower frequencies, they also sound as though they are coming from large animals.[14]

Not only do American Crows use specific sounds to communicate, they also package these sounds into series. Most students of crow language believe that it is both the patterning of calls and the type of call combined with the context that fully defines the crow's message. Like vocal smoke signals, crows may vary the number, type, rhythm, and repetition of individual calls to convey different meanings to those within earshot.[15]

Assembly calls are not structured into clear patterns. This eases the locating of stationary, visible predators and reflects the fact that many callers are screaming at the same time. The resulting din may further benefit the mob by confusing and annoying predators more than would a regular pattern of calls, thus convincing the predator to leave quickly. But in spite of the lack of structure, complex information can still be obtained from the

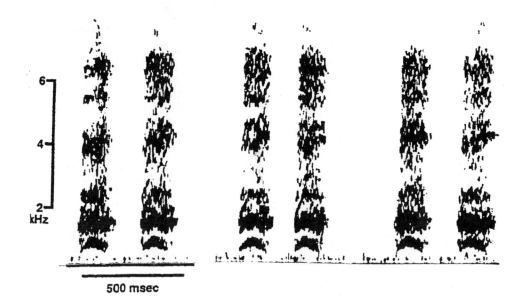

Six caws as given in real time by a wild American Crow illustrate the double-cadence rhythm used to claim territory (redrawn from Parr 1992)

sound of the mob. Mobbing tempo and intensity ebb and flow as predators freeze or move and as callers perceive different levels of risk. The regular progression from kos to long and harsh caws to squalls and screams and finally to koaws inform distant crows about the movements, proximity, and severity of the danger. Those being recruited may be able to determine the urgency of the mob by the pace of the calling; the faster the pace, the greater the risk, attracting others to the mob more quickly. Over time the human listener may even determine the object of the crow's mobbing by the intensity of their cries.[16]

American Crows are more refined when defending their territories, often precisely structuring call sequences for clarity. Crows give short-duration caws in a double cadence to call group members to arms in defense of the

Duetting pairs of territorial American Crows in vocal combat

territory. Neighboring groups often respond to these doublets so that "countercawing" vocal bouts ensue. Chases may erupt at territory boundaries if neither family backs down from the doubled vocal threat. Doubling calls may add emphasis to the message and distinguish an important call to action from normal background cawing, begging, and scolding.[17]

Territorial boundaries are also advertised and defended using caws of various lengths and rhythms. Structuring territorial calls by pausing at regular intervals affords the signaler a quiet break to listen for a response. Taking turns calling and listening may be more efficient than unstructured vocal free-for-alls for crows interested in maintaining territories and testing neighbors.

Structured cawing and countercawing share many features with the songs of other birds that we know and love. Their repeatability and use in long-distance communication to advertise territorial occupancy is a defining feature of most spring and summer birdsong. American Crows take this a step further and use their signaling to assemble their family group efficiently if battle erupts.

Crows do not sing to attract mates like other birds. Courtship is apparently done at close range using mixtures of soft sounds, bowing postures, and mutual caresses. Crow pair-bonds are likely formed over long periods in nonbreeding group settings without the vocal fanfare that accompanies pairing in most songbirds.

Carrion Crows coordinate their social partners with vocal cues. In this European species, mated pairs hold large territories, the defense of which is coordinated and encouraged by cawing duets. Duetting apparently stimulates coordinated and heightened attacks. Mates are more likely to increase their aggression and engage in "pincer tactics" after duetting than after solo cawing. Crows using pincer tactics surround an intruder, with the female on one side and her mate on the other.[18]

The complexity of a communication system such as that of the American Crow and Common Raven is rarely matched in the animal kingdom. Corvids use many distinct elements to convey a variety of meanings. They blend some elements to modify meaning and convey emotion. The caller's identity is advertised. Visual display is used to enhance a vocal message's meaning. Basic call elements are composed into series, and calls and songs are passed culturally through generations. All of these attributes are more like human language than most other animal communication systems. Dogs, whales, and the great apes have similarly complex languages—and they share social lifestyles. A high level of social activity is fostered by the ability to perceive the subtleties of sound, a capacity that corvids, cetaceans, canids, apes, and people have all employed successfully. The cawing of crows allows many scattered family members and neighbors to coordinate their actions and learn efficiently from each other's experiences, good and bad.

A social lifestyle, or sociality, may account for why people have been so fascinated with dogs, crows, and apes during our own history. We relate better to other social species because we have solved nature's challenges in similar ways. We understand similar species more easily than those living alone or communicating in more foreign ways. Although we have gained considerable insights into how bees and ants communicate to organize their societies, perhaps in part because of their sociality, abundance, and economic importance, we have a harder time relating to less social and nonvocal species like electric eels. It seems also that we are more likely to coevolve culturally with similar species.[19]

As we coevolve with other species, not only do we become better at reading their actions, they become better at reading us. Dogs, for example, are experts at deciphering our gazes, hand signals, and vocal commands to find food. Their long period of close association with people has made them better at interpreting our signals than are either wolves or apes. Crows also appropriately interpret our gazes, strides, and vocal signals. Stare at a crow, and it will become alert and likely move away. Walk quickly past, and it will ignore you. Develop a regular ritual of movements before you put food in your bird feeder and crows will anticipate and move toward the feeder when you begin your "feeding sequence." Ravens are adept at following the gaze of people and probably many other animals they encounter. Perhaps corvids that regularly interact with people routinely monitor our nonvocal cues, following our gazes toward hidden food or danger, just as a dog does today.[20]

Crow communications open a door into the crow's mind. The complex patterning of calls, for instance, suggests that crows are mathematically inclined. Counting and distinguishing among messages that vary only in the number of calls may be a regular part of a crow's life. After listening to hundreds of crow calling bouts, Nick Thompson, a psychologist studying crow language, suggested that crows recognize and use a simple counting system with a lower limit of one and upper limit of six to code and decode vocal messages. Crows appear to count only whole numbers, not fractions. The type of call, its intonation, and whether one, two, three, four, five, or six are packaged together may all be used to encode information. One wonders, however, if some fraction is communicated if one of three calls is given with less intensity, on another pitch, or with shorter duration.[21]

This reliance on math to communicate may explain why corvids appear to have formidable math skills. Corvids and a few other birds are able to solve problems requiring an ability to count to around six. Otto Koehler, a German animal behaviorist, demonstrated this in a fascinating series of experiments involving captive Western Jackdaws. When presented with boxes covering food items, the birds quickly learned to stop turning over boxes after a specified number of items were obtained. This worked with up to six food rewards. For example, after being trained to retrieve five bits of food, a jackdaw would approach an arrangement of boxes and turn over boxes until it secured five tidbits. When it got the fifth reward, it returned to its cage and did not turn over more boxes. Western Jackdaws easily learned this task and even could associate several box colors with specific numerical tasks. In some experiments they learned to open green boxes until two rewards were obtained while also opening white boxes until five rewards were obtained. That jackdaws mentally kept track of encounters was demonstrated in one of Koehler's experiments in which a bird trained to get five food items turned over boxes that revealed one, two, one, and zero items in succession. After obtaining these four rewards it returned to its cage, an apparent failure. But it quickly returned to the box line, sidled up to the original first box and bowed once, then went to the second box and bowed twice, and then bowed once in front of the third box. After these four bows, which seemed to represent a mental recounting of the previously obtained rewards, the bird went to the fifth box, flipped it over, and got the last tidbit.[22]

A counting Western Jackdaw

The crow's use of rudimentary language raises another possibility: good talkers make good liars. The fact that crows and ravens are so often seen as tricksters and outright prevaricators in our legends and stories suggests that early people were often deceived by cunning corvids. Perhaps our ancestors just did not accurately decode the signals given by their feathered companions, or is it possible that the corvids actively deceived? Ravens are capable of deception, albeit in a nonvocal manner. When a raven caches surplus food in the presence of other ravens, it often "lies" about the cache's location by falsely appearing to cache in one location, then actually caching elsewhere. The deceptive bird goes through the motions of placing a morsel into the ground but, at the last moment, hides the food in its beak and leaves the site to cache elsewhere. Sneaky crows have likewise deceived us. John Withey climbed a tall Douglas-fir to the site of a crow's nest that Marzluff had watched the crows build and tend on several occasions. The parent crows scolded us, as they typically do when we go to band their young. As we neared the nest site, we heard begging, yet it was coming, not from the nest the crows had previously "shown" us, but from a nearby Western Red Cedar. Even though we watched this pair of crows many times, we never saw them approach their actual nest. They always faked us out by going to the other "nest site." Steller's Jays (*Cyanocitta stelleri*) use a similar strategy by going to a location quite distant from their actual nest to confront and scold inquisitive crows. The crows apparently conclude that there must be a nest nearby and search fruitlessly, in the company of the scolding jays, before departing the area.[23]

We know that vocal deception occurs in some birds. Two tropical forest birds, the White-winged Shrike-tanager (*Lanio versicolor*) and Bluish-slate Antshrike (*Thamnomanes schistogynus*), regularly practice deception by giving hawk alarm calls when hawks are not present. The reason for their deception is clear; distracting the flock of birds they are with allows them to grab valuable foods. These birds are usually sentinels for foraging flocks of various tropical bird species, and they "cry wolf" when others are pursuing insects. We suspect that a similar form of deceit may also occur among a variety of smaller corvids, notably jays, who routinely mimic hawk cries. A hungry jay, seeing another bird with food, may cry like a hawk to startle the bird into dropping its prize, which the tricky jay then seizes.[24]

In captivity, ravens and crows label people with unique calls, like Lorenz's raven, who labeled him Roah, but in the wild this is unknown. Perhaps corvid language is sufficiently complex that existing calls are sufficient to distinguish among dangerous and passive human situations, familiar and strange people, or natural and human-made foods. Maybe standard calls are uttered in slightly altered ways to encode the human element. Or perhaps there is a crow call for our species that is modified to tell other crows whether we are friendly, neutral, suspicious, or dangerous. Controlled experiments are needed to sort among these possibilities. We anticipate that they will eventually show crow language to be acutely tuned in to people, but we do not presume that we are honored by specific crow calls in the manner that we honor them with specific words.

The glimpses we have into corvid vocalizations, and the promise that further knowledge holds to tell us more about communication, culture, and counting in another species, will certainly tempt others to study cawing crows. We look forward to these discoveries. Our current knowledge is really only utilitarian. We know some basic words and phrases. It is as though we are traveling in a foreign country without speaking the language; we miss many cultural nuances. Until we can speak crow, we will not fully know the crow.

SEVEN

Reaping What We Sow

The seemingly inexhaustible herds of Bison (*Bison bison*) that churned across North America's central plains were essentially extinguished before the last quarter of the nineteenth century. The combined impacts of wanton slaughter and hide hunting had taken a terrible toll. At the same time, midwestern states like Michigan, Ohio, Indiana, Illinois, and Wisconsin had their rich forests felled for lumber and their fertile prairies plowed and planted with corn, wheat, sorghum, and other grains. As America's population expanded, so did the exploitation of the timber resources; by 1920, just 4 percent of the eastern forest remained uncut.[1]

Corn is high on the menu of the American Crow

These sudden and vast changes in the landscape signaled an end to the raven's domination. And a start to the crow's.

Common Ravens disappeared as nesting birds in southern Michigan after only a decade of land conversion (1860–1870), and the last raven in Nebraska was seen in 1936. American Crows marched swiftly and steadily west and north as agriculture replaced prairie and forest. They were first spotted in southern Michigan in 1858, but by 1892 they were present throughout this region as pioneering populations advanced into northern Michigan. Through the first third of the twentieth century, crows were rare in the interior of the western United States. By midcentury, American Crows were the most common birds among the region's farms, streams, and settlements. Canadian physician and ornithologist C. Stuart Houston documented the same changes on the prairies of Saskatchewan. Ravens were common associates of Bison through the early 1800s. Crows were seen only in small summer flocks. Crows multiplied and ravens retreated as Bison declined, agriculture increased, and trees invaded the fire-suppressed, urbanized prairies.[2]

This changing of the corvid guard throughout much of North America illustrates our pervasive influence on crows and ravens. As settlers shot ravens, reduced the forests they used for nesting, and destroyed their traditional, reliable food sources like dead bison, ravens retreated. They moved to remote forest remnants and cliff-studded arid lands, typically in the north and around the western mountains. In contrast, we increased suitable crow nesting habitat by snuffing fires and planting trees on the prairies and fragmenting and thinning the forests. We converted prairies and forests to croplands on a massive scale. Crow populations responded by increasing and reaching out to colonize the entire continent. Crows also underwent a major cultural shift. To tap agricultural riches in the cold north, they began migrations from northern breeding grounds to southern wintering areas. Others remained as year-round residents in mild coastal or southern areas.[3]

A variety of evolved strategies may enable crows to exploit people more than is possible for ravens. Erik Neatherlin, a skilled climber, radio tracker, and raven trapper, studied ravens and crows on Washington's Olympic Peninsula with John Marzluff. Together they spent seven years and radio-tagged more than a hundred birds to reveal how Common Ravens and American Crows used people in this remote, temperate rainforest. Crows respond most strongly to the presence of people. They are eight times more abundant within six-tenths of a mile (one kilometer) of people than they are further from people. Ravens, in contrast, are only slightly more abundant near people. Crows benefit from people by doubling their reproduction and enhancing their survival. Their mobility allows them to make daily excursions of up to nineteen miles (thirty kilometers) to reliable but distant food sources, like garbage dumps, larger settlements, and sympathetic residents when local resources become scarce. Tolerance is also important for crows. They defend only the immediate nest area. Crow populations can quickly find and colonize new campgrounds and remote settlements, rapidly increase population size at such sites, and attain high densities by tolerating close neighbors. Ravens are intolerant of close neighbors. Despite efficiently exploiting distant resources and reproducing at a high rate near people, ravens cannot attain densities anywhere near that achieved by crows because pairs aggressively defend most of their large home ranges from other neighboring pairs. A social nature allows adjacent nesting crows to share half of their home ranges. In contrast, less social neighboring Common Ravens shared only one-sixth as much of their territories.

Crows' abilities to survive on small territories and fly far to find new havens allow them to track changing human resources. But we think crows' unparalleled ability to reap nearly everything people sow is the fundamental reason for their recent success. Crows are ingenious foragers who eat a huge assortment of foods, quickly learn to exploit new entrées, use tools under some conditions, and store surpluses for lean times. These flexible, resourceful ways serve them well in the rapidly changing world people create. Their opportunistic foraging, exacerbated by a marked fondness for corn, allowed crows to exploit our increasingly agrarian lifestyle in North America. But for all the good that crows reaped from their catholic diet and clever ways, they also reaped an equal amount of persecution and hatred. It is this intersection between fostering crow populations with supplemental food on one hand and attempting to control burgeoning crow numbers on the other that allows us to most clearly document the cultural coevolution between crows and people.

Crows eat just about everything. In the most comprehensive study of crow diets ever done, E. R. Kalmbach of the U.S. Biological Survey collected the stomachs from 2,118 crows—1,340 adults and 778 nestlings from forty states and several Canadian provinces—and identified 650 different food items. At one point in time, Marzluff's pantry, refrigerator, and freezer only had 290 types of food, beer included. Kalmbach studied crow diets in mostly agricultural areas in the early 1900s and determined that just over a quarter of the annual adult crow diet is composed of animal matter, chiefly insects, including many important agricultural pests. Others have confirmed this, noting that crows eat European Corn Borers (*Ostrinia nubilalis*) and Gypsy Moths (*Lymantria dispar*). Consumption of corn borers by crows lowers this insect pest's over-winter survival yet does not damage growing corn, because the bugs are consumed during winter by migrant crows. Even a single crow can eat a lot of insects. Kalmbach noted three crows whose stomachs included 85 May beetles, 72 wireworms, and 123 grasshoppers. A brood of four nestlings was fed 418 grasshoppers during the few hours before the birds were collected. Nestling crow diets are four-fifths animal matter, mostly insects.

An adult crow's diet is another matter. Kalmbach's study showed that the mature crows' menu was made up of more than half grain, mostly corn. This is typical for crows in agricultural areas, but diet varies widely among study locations. On Washington's Olympic Peninsula, adult crows ate no grain, and nearly two-thirds of their diet consisted of animals. In Florida, Ottawa, Wisconsin, New York, and Washington, some truly bizarre food items have been documented. Our favorites include: long-horned beetles, large eel-like salamanders, and walking catfish caught by crows or stolen from River Otters (*Lontra canadensis*) by teams of crows, one of which pinches the otter's tail while the others grab the fish. They also eat River Otter dung, emerging caddisflies, adult House Sparrows (*Passer domesticus*) and European Starlings (*Sturnus vulgaris*), bats, snakes, mice, and possibly even small pigs, Raccoons, and deer. Yes, crows have been seen attacking young and weak pigs and deer. One crow rode a piglet for twenty minutes, continuously pecking at it, before giving up. The prize for gross food goes to a Seattle crow that pecked dried human vomit off the side of a downtown building.[5]

You would not think of a crow as a picky eater, but in fact they can be exceedingly selective foragers. In Manitoba, for example, crows are attracted to emerging garter snakes in the spring. The snakes are superabundant, and the crows eat only their livers, deftly opening a one-and-a-half-inch-long section (four centimeters) of the dispatched snake's skin. Eating liver may give

Crows team up to steal an otter's meal

crows needed vitamins like A, B, and D or metals like iron and copper, but it can also expose them to a variety of environmental contaminants like arsenic. Short-term concentrations of food appear to drive this unusual behavior. In Seattle, where garter snakes are usually encountered singly, crows eat them whole. In Ontario, crows sometimes eat only the heads of earthworms, leaving the rest of the bodies scattered about the lawn. Looking at the anatomy of an earthworm makes sense of this selectivity—earthworm bodies are mostly intestine and waste; heads house the eggs, sperm, brains, and gizzards. Oddly, in such times of plenty, crows do not store uneaten food for later use. Apparently snakes and worms just don't keep well.[6]

Today's crows can take or leave our grain. Exploitation of superabundant grain has likely been a key in their ability to exploit people. Crows' shift to a corn diet allowed them to colonize the Midwest and spread widely across North America. Recently, though, they have shifted their diet further to exploit superabundant resources generated in our cities. In Seattle, more than half of a crow's diet is human refuse. Insects and worms, other small animals, and fruit round out the urban crow's diet. This is unquestionably a fundamental and recent cultural shift. Crows on the nearby but less settled Olympic Peninsula have a more traditional diet, about one-third refuse, over half animal matter, and a bit of fruit. Social learning likely allows American Crows routinely and rapidly to adjust their diet to exploit food provided, directly or indirectly, by nearby people.[7]

The crow's ability to learn exactly what to eat in a human-dominated environment can be impressive, even to a child. Eight-year-old Zoe Marzluff proved this by offering crows a choice of fried potatoes in two types of bags. One bag was from a McDonald's restaurant, and the other was a plain brown bag. Crows consistently approached the McDonald's bag before the plain bag, even though both were within a yard (meter) of each other, of the same size, and contained the same food. Crows had apparently learned to associate the restaurant's logo with a tasty snack. Groups of crows crowded around the bags as one especially brave, hungry, or knowledgeable crow ripped one open. The feeding environment was thus ripe for social learning facilitated by simple observation. Any young crow or new immigrant had only to watch what the experienced crows ate to learn the local customs. Cultural evolution is easy when a curious, social, and intelligent animal meets a rich and variable environment.

It may take patience and vigilance, not to mention brains and a cast-iron stomach, to scavenge effectively from people. If crows and ravens have taught us anything as we try to study them, it is to be patient. They will wait for hours or even days before eating new foods or using new feeding locations involving our traps. This patience likely keeps them out of many deadly situations. When they finally decide to eat, one gets the impression that they are always ready to spring away from some unseen danger. Tense and fearful ravens typically jump up as if their toes were pinched, performing what we call "jumping jacks" as they approach new foods. American Crows spend more time looking for danger and less time eating in environments rife with human disturbance than in less disturbed environments. But once the danger of a new site or food is overcome, crows and ravens quickly get used to the new food and take full advantage of it. They habituate to disturbance and novelty more quickly than most species, which may allow them to outcompete powerful scavengers, like eagles or vultures, in places people frequent. On the streams of the Pacific Northwest, for example, American Crows and Common Ravens beat Bald Eagles to decaying salmon carcasses because they quickly get used to fishers, rafters, and nature watchers.[8]

Feeding crows illustrate how their communication, recognition, and learning abilities can be used to coordinate their flocks. Social vertebrates like people, wolves, and crows possess all these abilities and use them to adapt to rapidly changing social settings. Northwestern Crows, for example, forage in groups of upwards of a hundred birds on the rocky shores of Washington. While some are busy looking among the seaweed, shells, and rocks for food, others are vigilant for predators and easy meals. These "scroungers" approach others who have food and either aggressively or passively try to steal it. Even thievery, however, has rules in crow society. As Renee Robinette, a research associate at the University of Washington, has discovered, scrounging involves chasing, biting, and hitting only when it involves unrelated crows. Relatives use the more civil, passive sidling and take food quietly from one another. Northwestern Crows use individual recognition, familiarity, and kinship to maintain order and efficiency where chaos could easily rule. Natural selection has shaped social behavior so that large flocks can forage efficiently on one hand and keep from becoming prey themselves on the other.[9]

Caching, or storing surplus food for later use, is a defining characteristic of corvids. The seed-caching abilities of the smaller corvids, Pinyon Jays, European Nutcrackers (*Nucifraga caryocatactes*), and Clark's Nutcrackers, are truly remarkable. The birds spend most autumn days collecting durable, hard-shelled pine or hazel seeds. Their bills are specialized chisels, making it easy to access seeds within tough cones. Their throats are expandable; nutcrackers have specialized storage pouches to carry tens to hundreds of seeds at a time to caching grounds miles (kilometers) away. Each autumn tens of thousands of seeds are cached singly or in small clumps by each bird. Amazingly, these birds use their powers of memory to find each cache's location so that the seeds can be recovered throughout the winter and following summer. Nutcrackers can accurately dig down through a yard (meter) of snow to find their buried seeds. One is tempted to surmise that if a 3.5- to 6-ounce (100- to 170-gram) bird can do that, surely a 14-ounce (400 gram) crow or a to 4-pound (2 kilogram) raven can do better.[10]

Actually, although all species of crows and ravens cache prodigiously, they usually store perishable items for relatively short periods. American

Clark's Nutcracker (a) preparing cache site, (b) retrieving seed from throat pouch, and (c) caching a pine seed

Crows cache a variety of animals too large to be consumed in one sitting. Northwestern Crows cache clams for a few days. Common Ravens in the Arctic cache bird eggs. Common Ravens and American Crows at a large mammal carcass in winter fly continuously from the feast with mouths over-flowing with fat and meat. In this way they scatter bits and pieces of the mammal throughout the forest, but usually within a few hundred yards (meters) of the kill site. The perishable meat may stay frozen in the snow or ground for weeks or months. It may elude other corvids or scavenging mammals. The meat could, therefore, sustain a crow or raven for an entire winter. But it does not. Few of these caches are ever revisited or recovered by the cacher. Long ago the crows have moved to another fresh kill, an agricultural field, a bird feeder, a garbage dump, or a fast-food restaurant dumpster in a nearby town.

Can crows and ravens be forced to focus on their caches and reveal their true memory abilities? As a doctoral student in the University of Washington's psychology department, Bob Reineke attempted to do just that, testing American Crows in a large aviary containing eleven hundred closely spaced cache sites of sand-filled film containers. Reineke's crows worked for dry dog food. Although they are not as strongly motivated to cache as jays and nutcrackers, the crows cached after an initial bout of eating but while still hungry. Four times out of five, they could recover their caches, even if forced to wait thirty days. Furthermore, in a given recovery session, they remembered which caches they had emptied and which they had missed in the previous session. Not bad for a generalist that rarely returns to caches in nature more than two to three days after making them. In similar tests, the specialist jays and nutcrackers find their caches nine times out of ten.[12]

For corvids, Common Ravens may be at the intellectual bottom of the barrel when it comes to cache-site memory. Bernd Heinrich studied Common Raven caching in a large outdoor aviary he and Marzluff built in the Maine woods. Marzluff himself needed to cache doughnuts for sustenance during the aviary building. Ravens are also expert doughnut cachers, but they usually cached meat provided by Heinrich and his colleagues. Heinrich's ravens did not accurately recover their caches more than a few days after they were made. How can the world's arguably smartest bird forget where it hides its caches? Was the raven given an inappropriate test? One might suggest that Heinrich's birds were so smart that they knew it didn't matter if they forgot their caches, since other food would soon be provided. Or perhaps they knew where the caches were but chose not to recover them. Possible, but doubtful. It seems more likely that ravens do not remember their cache locations for long because, in nature, they rarely have the opportunity or reason to recover their caches after a few days. Food cached by ravens often rots, is stolen, is far away at a past feeding location, or is rendered obsolete by the next food item the ravens find.[13]

Ravens do not need to remember where they cache for long. But this does not mean that caching is unimportant to them. Just watch one cache. A chunk of meat is surreptitiously held in the beak as the bird sneaks away from the foraging gang. The food is placed in the ground with a sharp stab of the powerful bill. It is carefully covered with grass, sticks, a pinecone, dung, or snow. More often than not, it is moved later or actually not deposited in the carefully reconstructed site—a raven's version of a magician's "sleight of hand." Ravens and other corvids fake caching a lot. This confusing hiding behavior presumably keeps other ravens in the area uncertain of the location of the cache. Caching ravens are thus deceitful and caches are important, at least in the short term. Why else would ravens lie about a cache's location? Heinrich and his colleagues demonstrated that ravens in a large aviary watch their flock members cache and try to pilfer the caches. These are clearly smart birds that remember where caches are and use this knowledge to retrieve them if and when they are needed. They just do not need them very often. Ravens may know when perishable items are no longer worth recovering. Or maybe for ravens it really is just a shell game.[14]

Caching teaches us a lot about natural selection. Crows and ravens do not rely on their caches for sustenance to the extent that Pinyon Jays and nutcrackers do. Pinyon Jays and nutcrackers that forget where their seeds are buried may die over the winter or at the least will produce fewer offspring the following breeding season. Crows and ravens can forget where their caches are and still find other sustenance for survival and reproduction because of their extremely catholic tastes, wandering ways, and problem-solving abilities. Accordingly, their powers of spatial memory are not entirely focused on cache recovery. Rather, crows and ravens are adapted by natural selection to changing conditions. This has given them the ability to solve a wide range of unanticipated problems with individual and social learning, teaching, and insight.

Caching also shows us that the behavior of crows and ravens is not always obviously under the control of natural selection. Rather, the fact that many corvids seem possessed to cache all sorts of shiny objects causes us to ponder whether crows and ravens enjoy beauty and seek novelty. In the late 1800s, the famous Canadian naturalist Ernest Thompson Seton studied an old, wild crow with a patch of silver feathers near his beak. "Silverspot," as Seton called him, stored a hatful of white pebbles, clamshells, and bits of tin in a carefully guarded subterranean cache. Silverspot routinely added objects to the cache and admired his booty. Seton believed that the handle of a white china cup was the gem of the collection. Silverspot was very attentive to his cache and moved it to an unknown location soon after Seton cataloged the contents. Perhaps this cache advertised Silverspot's prowess to other crows, or perhaps it simply illustrates that aesthetics can motivate crow behavior.[15]

The caching antics of crows and ravens often cause mischief. A raven living along a Minnesota golf course was infamous for stealing balls before the golfers could hit their next shot. This led to an additional bird term in golf parlance: "a raven," which was, in contrast to a "birdie" or an "eagle," terrible! Ravens in the Mojave Desert are also infamous for stealing golf balls, and they cache them in large piles near their nests. A wild crow gained notoriety in 1997 for stealing a $450 gold bracelet from a Florida man. It was found two weeks later and thirty-five miles away and returned to its rightful owner. These stories of stealing and caching strange objects leads us to hypothesize that the corvid preoccupation with shiny objects is indeed a form of self-advertisement. If other family or flock members are impressed by the valuables, then the owner's social status might increase.[16]

Next to crop raiding, the habit that most often gives crows and ravens a bad name is their fondness for other birds' eggs and nestlings. Few things are as memorable as watching a crow shred a favorite songbird's nest, shamelessly gulping down defenseless nestlings as the parent birds shriek hysterically and fly at the black grim reaper. Unless, of course, it is a huge raven systematically killing an entire brood of helpless and clumsy ducklings. Though graphic and real, are these acts typical of foraging corvids? Do such acts lower the viability of songbird populations? Are crows and ravens really *vicious* predators?

Nearly every corvid is known to prey on the eggs and nestlings of other birds. We know that corvids are extremely adept at finding birds' nests, whether on purpose or by chance. Nests within 765 yards (700 meters) of American Crow nests or those especially exposed are quickly dispatched. Hooded Crows pull out all the stops to find nests. They watch parent loons, track bird researchers, and remember nest sites from one year to the next. Geir Sonerud and Per Fjeld, behavioral ecologists at the Agricultural University of Norway, observed a tagged, breeding male crow in Norway recheck previously preyed-on nest sites within a breeding season and preferentially recheck new nests placed at the site of a previous year's nest. Their results demonstrate that a single nesting Hooded Crow can reduce the hatching success of other birds nesting on its territory. Still, demonstrating an ability to find and prey on nests efficiently does not necessarily mean that crows or ravens limit other bird *populations*.[17]

Common Raven preying on ducklings

Corvids love eggs. If given a choice, American Crows will take small, easily carried eggs before they take large ones. Usually only a single egg is carried at a time, but once a nest is found, the discoverer will return to empty it, caching each egg safely in the ground well away from the nest before returning for the next prize. Where bird nests are clumped into large colonies, a single corvid that caches as well as eats eggs can remove a lot of eggs. Ed Murphy, a raven and seabird researcher at the University of Alaska in Fairbanks, and his graduate students have observed Common Ravens preying on eggs of Common Murres (*Uria aalge*) at a colony in western Alaska. Typically, a raven will land near an incubating murre, pull it off its egg and then fly with the intact egg in its bill inland from the cliffs. Ravens scatter their caches up to two and a half miles (four kilometers) inland across the tundra. Each egg is placed in a single subterranean hold, where the permafrost refrigerates the booty throughout the summer and early fall. While a raven is preying on and caching murre eggs it may take an egg every couple of minutes. Paul Rossow, a graduate student working with Murphy, estimated that a pair of ravens preyed on and cached about two hundred eggs each season. Jay Schauer, another student working with Murphy, showed that ravens preyed primarily on freshly laid eggs of murres that had no breeding neighbors at the time. Murres do not build nests, and they breed literally shoulder to shoulder. Once they begin incubating, they aggressively defend their eggs against ravens and gulls, but successful defense seems to require the presence and defense by two or more neighboring incubating adults. Perhaps in this way, the culture of egg harvest by ravens reinforces the culture of coloniality and synchronized breeding by murres.[18]

Fondness for eggs of all types appears to be innate. As a postdoctoral student in Maine with Bernd Heinrich, Marzluff and his wife, Colleen, raised broods of ravens and offered them a variety of objects. The ravens never liked getting squashed opossum pushed down their throats. But young ravens, barely able to fly, gobbled down any egg left in their cage. This included warbler eggs, chicken eggs, plastic Easter eggs, and even Heinrich's prized ostrich egg, which was quickly removed before the birds broke its thick shell. Thus, it is not surprising that in the wild, corvids are mostly associated with predation on eggs, rather than nestlings, which are more commonly preyed on by small mammals.[19]

Occasionally, corvid nest predation can be devastating. American Robins (*Turdus migratorius*) in some yards may never fledge young, primarily because of American Crow predation. Herring Gull (*Larus argentatus*) and Ring-billed Gull (*Larus delawarensis*) nesting colonies can suffer nearly complete nesting failure at the beaks of American Crows and Common Ravens. African Paradise Flycatchers (*Terpsiphone viridis*) are declining in Dar es Salaam, Tanzania, in part because of increasing House Crows. However, most documentation of the extent of nest predation by corvids shows they are only one of many nest predators and destroy relatively few nests. Crow stomachs investigated by Kalmbach revealed that wild birds and their eggs only accounted for 0.3 percent of adult and 1.5 percent of nestling crows' annual diets. On the Olympic Peninsula, Steller's Jays are important nest predators, but American Crows and Common Ravens are minor predators. Marzluff and his students watched breeding adult American Crows forage on 435 occasions and saw them prey on only three songbird nests. They also watched a thousand artificial nests, made to emulate those of the threatened Marbled Murrelet (*Brachyramphus marmoratus*) and placed high in the old-growth forest canopy. American Crows and Common Ravens preyed on only 5.7 percent of these nests. Moreover, these corvids were only two of the fifteen nest predators documented—Flying Squirrels (*Glaucomys sabrinus*), arboreal White-footed Mice (*Peromyscus keeni*), and Steller's Jays were more important. Indeed, the diversity of nest predators explains why removing American Crows from some duck nesting grounds does not guarantee duck nesting success. Where crows are removed, skunks and other nest predators can step in and eat more eggs, compensating for the lack of corvids. The inability of crows to affect populations of some birds, even with substantial nest predation, is seen daily in the United States, where, despite losing half of their nests to predators each year, American Robins never seem to be in any shortage.[20]

Crows and ravens often get blamed for more nest predation than they deserve simply because of their conspicuousness. In our studies on the Olympic Peninsula, the abundance of crows and ravens in forests near human camps and settlements correlated with the risk of nest predation, even though the birds actually preyed on few nests. The problem might be guilt by association, for crows and ravens often indicate areas where many predators congregate. In our case, this included concentrations of important predators such as jays, squirrels, and mice. In other situations it may include rats, feral and domestic cats, Raccoons, opossums, and snakes. These mostly nocturnal animals are important nest predators that, unlike the daytime, highly visible crows, are rarely caught in the act by casual observers. Recent research using video cameras to monitor songbird nests, however, routinely documents the predatory ways of small mammals, jays, snakes, and hawks. People may be more likely to see a big, black, animated crow or raven in the act of nest predation, but crows and ravens are more often mere harbingers of death rather than agents of actual destruction.[21]

An exception is in the increasingly human-dominated landscape of the Mojave Desert, where Common Raven populations have increased 1500 percent from the 1970s to 2000. Adult ravens nest on power poles and juveniles congregate en masse at landfills and other human developments. The desert is also home to the rare Desert Tortoise (*Gopherus agassizii*), and even though ravens cannot kill and eat the adult tortoises, young tortoises have relatively soft shells that are easily punctured by powerful raven beaks. Piles of pecked tortoise shells are found in some raven nesting territories, but the full impact of ravens on tortoise populations was not revealed until Bill Kristan and Bill Boarman placed one hundred Styrofoam tortoise models throughout a 190,000-acre (770-square-kilometer) section of Edwards Air Force Base, California. Ravens attacked nearly one-third of the models during a brief four-day period of exposure. The researchers also saw that the more ravens there were, the greater was the risk of predation. Ravens attacked any tortoise model near landfills where nonbreeding ravens congregated and near successful raven nests. The magnitude of predation and the wide extent of high-risk areas suggest that ravens are a serious limiting factor to Desert Tortoise populations. Raven increases in response to people threaten tortoise reproduction. But people also directly reduce Desert Tortoise populations by running over tortoises with cars and taking them home as pets. Our cumulative activities have caused this once common species to be listed as "threatened" under the Endangered Species Act.[22]

In a perfect world, the diversity of nature needs predators. They are a natural part of the Earth's ecosystems, filling important roles. Predation may not be pretty, but it helps keep prey populations healthy and in check. Preda-

Common Raven sizing up a Desert Tortoise

tors can provide important resources for other species. Old crow and raven nests are routinely used by a variety of hawks and owls. Predators can also promote greater diversity than is possible in predator-free areas by keeping dominant competitors from overrunning others. Without predatory starfish, the tide pools of the Pacific Northwest would be clogged with only a few species of mussels. But with starfish, potentially dominant mussels are kept in check, and a diversity of sponges, nudibranchs, chitons, and other shellfish share the rocks.[23]

In our increasingly human-dominated world, the relative species composition or balance of nature has changed. This is shown clearly by worldwide increases in many corvids today. Even though corvids are just a few of the many nest predators, they are having a detrimental impact on some species. Species that are rare because of other human activities, like Desert Tortoises, California Least Terns (*Sterna antiallarum browni*), Mountain Plovers (*Charadrius montanus*), Long-billed Curlews (*Numenius americanus*), Steller's Eiders (*Polysticta stelleri*), and Marbled Murrelets, may be pushed closer toward extinction through unnaturally high predation by corvids on their eggs, nestlings, and juveniles. We will discuss how people can manage these grave effects in the next chapter.[24]

Crows employ an amazing diversity of behaviors to obtain food. They fish, probe, flake bark, sally after flying insects, dig, hammer, flip over rocks and cow pies, and use animate and inanimate aspects of their environment to enhance their food gathering abilities. Their foraging innovations often include making and using tools.

Consider how American Crows in Florida use our livestock. Some of these southern crows appear to clean lice and ticks from feral hogs and cows. To do so, they clamber about, over, and under sleeping piglets, walk on full-grown hogs, and hang from cows' tails. Lawrence Kilham, the astute crow biologist from Dartmouth, watched groups of up to three crows peck the heads, ears, backs, and bellies of sleeping piglets. The piglets seemed undisturbed even as the crows probed them up to sixty times each minute. The cows even seemed to facilitate cleaning by crows—one-third of the cows held their tails out to facilitate a thorough search. These associations between crows and farm animals appear to be an emerging symbiosis—an association where cooperation allows both species to benefit. We view it as an extended form of cultural coevolution, a change in crows' foraging culture in response to animals domesticated by people rather than in response to people themselves.[25]

Crows are nothing if not ingenious in their ways of getting a good meal.

Crows glean lice from pigs' skin

An obvious expression of ingenuity is their use of gravity to help open hard-shelled prey. American Crows, Northwestern Crows, Hooded Crows, Rooks, Western Jackdaws, Carrion Crows, New Caledonian Crows, Common Ravens, and Chihuahuan Ravens drop nuts, clams, crabs, bones, and even turtles on hard surfaces to crack them open. Fish Crows and Common Ravens even drop sticks and dirt clods on nesting gulls to flush them from their valuable and tasty clutches of eggs.[26]

The dropping of food items has been studied in detail and illustrates the sophistication and learning employed by crows and the powers of natural selection to shape animal behavior. American Crows adjust the height at which they drop walnuts in accordance with shell thickness, substrate hardness, and the nearness of other crows. To crack walnuts and other hard prey, crows must fly rapidly skyward, which takes a lot of energy, drop the nut, watch it hit, and fly down to retrieve it. Walnuts are hard to crack, so crows must drop them as many as fifty times before they finally yield their meat. Rather than fly to the maximum height for each successive drop of a nut, crows save energy by reducing successive drop heights. Crows started dropping English walnuts from three yards (meters), but after four drops flew only half as high. They also saved energy by dropping thin-shelled English walnuts from lower heights than hard-shelled black walnuts, and by dropping walnuts from lower heights onto hard asphalt than onto soft soil. They increased their chances of enjoying the fruits of their labor by dropping walnuts from lower heights when other crows were nearby and anxious to steal cracked nuts than when alone.[27]

In Japan, the culture of nut-cracking by Carrion Crows is getting a boost from our increasing use of automobiles. Rather than relying only on gravity to crack hard-shelled walnuts, ingenious Carrion Crows in Sendai, northern Japan, place seeds on roadways and wait for automobiles to run over them. Instructors at the Kadan driving school noticed crows dropping walnuts on roads of the driving course in the 1960s. But it was not until 1975 that researchers like Hito Higuchi reported the first apparently purposeful placement of walnuts in front of cars. Photographs by Higuchi and his colleagues clearly show crows first waiting with walnuts for traffic to stop at an intersection, then flying down in front of stopped cars to position nuts in front of tires, and finally returning once the cars have moved on to eat the cracked nuts. This behavior has spread slowly outward several miles (kilometers) from the driving school over a twenty-year period and is increasing to this day. This slow spread likely represents cultural transmission of a behavior initially learned through trial and error. The innovating crows' offspring and neighbors would certainly observe and could easily acquire the nut-cracking habit. If social learning allowed the behavior to spread, then nut-cracking would be a cultural trait. Nut-cracking is also being facilitated by cultural changes in Kadan's drivers—a clear case of cultural coevolution. People help the crows by intentionally driving over nuts placed on roadways. Higuchi tells us that people enjoy driving over nuts for crows and points out that this "attitude may be the reason why crows developed the spectacular behavior at Kadan."[28]

Indeed, a delicate interplay between crows and people may be needed to favor the use of cars as nutcrackers. Cars are not good nutcrackers when they are too infrequent to crack nuts open regularly or too frequent to allow crows safe access to open nuts. In Olso, Norway, Hooded Crows flock to the few walnut trees, pick nuts, and accurately drop them on roads from heights up to twenty-seven yards (twenty-five meters). Marzluff and Geir Sonerud watched these crows drop 102 nuts in the autumn of 2004. In no case was a car employed as a nutcracker. Cars regularly drove and parked in walnut drop zones. But there were no traffic signals to create the pulsed traffic flow

A Japanese Carrion Crow uses the automobile to crack walnuts. Developed from a series of photographs.

exploited by crows in Japan. So, the coevolution is meager in comparison to the highly coevolved system in Japan. Norwegian crows have expanded their diet to use walnuts, which were brought to Oslo, planted, and nurtured by people. Residents of Oslo are starting to notice the feeding behavior of crows. Perhaps with time, Norwegian crows will come to use automobiles as nutcrackers and Norwegian people will adjust their driving to encourage crows. For now, we can simply enjoy the cultural diversity of nutcracking crows around the world.

Nut-cracking by American Crows and Carrion Crows is sophisticated, but whelk-cracking by Northwestern Crows is an analytical masterpiece. Reto Zach, a postdoctoral researcher at the University of British Columbia, watched Northwestern Crows search for, carry, drop, and retrieve whelks, marine snails often referred to as *conchs,* on the intertidal reaches of Vancouver's rocky beaches. He discovered that crows dropped whelks only at traditional sites where the rocks were suitable for breaking the hard shells yet not too close to the water, where an errant bounce could carry the gastropod out of reach. He was surprised to see crows take only the largest whelks, ignoring smaller ones even if no large ones were to be found. He reasoned that crows must be selecting their whelks to maximize their energy return. It turned out that crows eating whelks expended about 550 calories per whelk, most of it during the four to five drops required to crack the shell. The effort repaid with interest by large whelks, each of which yielded 2,000 calories. Smaller whelks, whose lighter weight demanded more drops from greater heights, cost a crow 900 calories to open and repaid just 600 calories. A crow eating small whelks would eventually starve to death. Northwestern Crows get the most out of cracking large whelks by flying just high enough to crack them yet not one expensive flap higher. Zach calculated that the optimal height to drop a large whelk was sixteen feet (five meters). Northwestern Crows drop whelks from an average of seventeen feet (5.2 meters), within a wing flap of Zach's calculation.[29]

Whelk-cracking by Northwestern Crows is a cul-

Northwestern Crows will gauge the proper height from which to drop their shelled prey

A Hooded Crow retrieves an ice fisher's prize

tural tradition that involves several discrete learning tasks. Crows test whelks before they select one for dropping. They thus have learned to select a whelk of the correct size so as best to repay their energetic expenditure. They have also learned where and at what height to drop whelks. This complex learning allows crows to fine-tune their foraging behavior to a particular type of food. Efficient foraging allows crows to devote more time than most birds to other activities: parental care, play, observing, learning, social networking, loafing, exploring, and exploiting people. It is also an important reason why they are able to live long lives and raise many offspring. It is no wonder that when we supplement crow diets, their populations explode—their efficiency at using the extras we provide guarantees it.

Crows go well beyond the use of gravity to help procure foods. Hooded Crows in Scandinavia learned to use the tools of ice fishers to score free meals. Other crows make their own tools. American Crows craft splinters to skewer insects in crevices. This sort of tool use likely evolves in complexity as crows are rewarded for their innovations. In this way, American Crows in a laboratory learned to use matchsticks, first as probes and later as levers, to extract food more efficiently from their mechanical feeders. Another captive American Crow used a small plastic cup to gather water, carry it to his food bowl, and moisten his overly dry food when its human caretakers forgot to. But the hands-down champion corvid craftsman is the New Caledonian Crow. This crow fashions a variety of spears and hooks from vegetation to fish insects out of crevices, drops nuts to crack them open, and, in the lab, bends wire into hooks to raise otherwise unobtainable food buckets. You can almost see the wheels turning in their brains as you watch these crows try to hook a bucket handle at the bottom of an eight-inch-long plastic tube (twenty centimeters) using a straight piece of wire. The bucket cannot be retrieved until the wire is bent. So after a few times, the crow bends the end of the wire into a simple hook, reinserts the wire into the tube, successfully hooks the bucket handle, and hauls the treat out of the tube. Humans did not learn to make hooks until relatively late in our tool-making careers (less than 100,000 years ago). Still, these stunts pale in comparison to the New Caledonian Crow's manufacturing of "stepped-tools" from the leaves of the pandanus, or screwpine, tree.

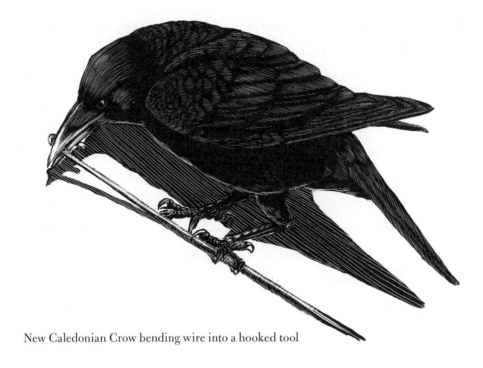

New Caledonian Crow bending wire into a hooked tool

Let's cruise to Grande Terre and Maré Island in the tropical southwest Pacific, where Gavin Hunt, a professor at the University of Aukland, and his colleagues have collected nearly four thousand stepped-tools. These tools are fashioned by ripping a sawblade-like section from a pandanus leaf that steps down from a wide base to a narrow point and keeps the naturally occurring barbs on one side facing away from the working end of the tool. These seemingly simple inventions are anything but simple. They are highly stylized and consistent in size and shape, indicating that they are made by applying a manufacturing rule—a combination of rip about two-tenths of an inch (four millimeters), crosscut one-tenth (two millimeters), rip, crosscut, and so on. People did not employ rules until we started making hand axes, some one and a half million to a hundred thousand years ago.

Moreover, the birds' tools are most often made from the left edge of a pandanus leaf. This is accomplished by "right-handed" birds. Right-handedness means that the right eye and left brain hemisphere are specialized for this tool-making behavior, just as the right eye–left hemisphere is specialized in other birds and mammals, including people, for other nonspatial sequential actions. The handedness that is associated with stepped-tool manufacturing implies that it is a complex, sequential motor task that benefits from efficient neural programming. This is just like our fine motor skills and language, which are also sequential and complex and benefit from lateralization in our brain's right and left hemispheres. This complex tool manufacturing, use, and intricate neural control puts New Caledonian Crows on a par with gorillas, chimpanzees, and people. The ability to process information and use the environment like a primate is undoubtedly one of the reasons crows have been so successful in the wilds of New Caledonia, the cornfields of our breadbaskets, and the urban jungles of our cities.[30]

Crows and ravens get into a lot of trouble. Sometimes we invent better and better scarecrows or kill a few birds to keep them out of our prized gardens. In other cases, losses to crows and people are more extreme. Juergen Fritz, a shepherd in southwestern Germany, ended up abandoning a favorite grazing area after fifty to sixty Common Ravens from a nearby garbage dump killed nineteen of his sheep. The JR Tohoku baseball team in Sendai, Japan, lost a critical, close game when fifty crows descended into their outfield and distracted players. Cities like Singapore, Tokyo, Dar es Salaam, and Chatham, Ontario, have become the sites of full-blown "crow wars."[31]

Introduced House Crows have been targeted for elimination from Singapore by the local government for more than twenty years. By encouraging local gun clubs to kill thousands of crows each year— ninety thousand were shot in 2003—the House Crow population has been reduced from more than a hundred thousand in the 1980s to around thirty thousand in 2003. Originally, House Crows were viewed positively. People introduced them to nearby Peninsular Malaysia to control caterpillars and transported them directly to Singapore as ship stowaways or pets. Now they annoy residents and are suspected of carrying disease and reducing the diversity of native birds. Thirty to forty thousand crows will need to be killed each year for another decade to reduce conflicts with people.[32]

The story is the same in East Africa. House Crows were shipped there in 1891 at the request of the British governor of Zanzibar. The crows kept the governor's streets clean. But in barely a hundred years the crow population reached half a million in Dar es Salaam alone. They are spreading across Africa, threatening rare birds and annoying people. The House Crow's royal appointment is now met with deadly force. In the 1990s, more than a thousand were caught and killed each week for a year.[33]

Native but growing populations of Jungle Crows are vandalizing Tokyo cemeteries for the food offerings left for departed loved ones. They carry away baby prairie dogs from the Ueno Zoo and inflict bumps, bruises, and cuts on people as they walk unsuspectingly close to crow nests. From 1987 to 1999, 296 crow assaults were reported to government officials. The number of Jungle Crows in Tokyo has quadrupled since the early 1980s; in 2001 there were an estimated thirty-seven thousand of them. Mixed feelings always seem to follow crows: some people love, feed, and care for them, while others cower from or despise them. Public concern is so great that more than five hundred people attended one of the five workshops held to address the issue.[34]

Poor garbage removal is at the root of the problem. Tokyoites put their garbage in bags at communal drop-off areas, concentrating food resources in easy-to-puncture plastic bags. This garbage is a crow's dream, spawning a culture of scavenging in an otherwise forest-dwelling crow. To reduce scavenging, a few cities handed out more than fifty thousand nets to cover trash bags. Crows quickly learned to go under and through the nets. The Tokyo metropolitan government countered by spending 1.5 million yen, or fourteen thousand dollars, in 2001. It formed a committee from its governmental agencies and paid exterminators to destroy more than eight hundred nests,

Bold Jungle Crows challenge pedestrians on Tokyo's streets

thirty-six hundred chicks, and fifteen hundred eggs in 2000 and 2001. To-kyo's governor, Shintaro Ishihara, intends to "make crow-meat pies Tokyo's special dish."[35]

Local scientists offer a more balanced, albeit longer-term, solution to Tokyo's challenging crows. They call for reducing the availability of food scraps in the garbage by improving collection and containers. They are de-signing garbage bags that crows cannot see through or that are laced with chili, salt, and vinegar to dissuade crows, suggesting that commercial sources put food waste in tightly sealed boxes, educating the public, and hazing birds where needed. Scarecrows take on modern form as large-eyed bal-loons fly from public places to frighten crows.[36]

Tokyoites will ultimately decide how their culture will respond to and reshape the culture of Jungle Crows. If Jungle Crows respond like other crows, then we suspect that the more stable and lasting solution will involve reducing their human sources of food. Killing them will only select for new cultural innovations by crows. Indeed, control efforts already seem to be running into difficulties. In the summer of 2003, the reproductive success of inner-city crows increased, probably because the young that fledged from nests had to compete with a smaller number of older nonbreeding crows, since many had been previously captured and removed. Even if crows don't compensate for their losses, successful control is difficult to sustain. Pro-grams in Singapore and Dar es Salaam are also teaching us the challenges of controlling crow populations.

Chatham, Ontario, is another ground zero for modern conflicts be-tween crows and people. American Crows have established a huge commu-nal roost, causing the usual concerns about noise, disease, property dam-age, and scattered trash. The response of people to the crows was anything but ordinary.[37]

Mayor Bill Erickson responded to citizen concerns in 1998 by ordering a cull of Chatham's estimated thirty thousand to one hundred thousand wintering crows. According to the *Daily News,* the local newspaper, the mayor suggested a small killing to scare the others away. But in 1998, the crows left Chatham before killing was necessary. In 1999, the plan was resurrected with the return of crows to their now-traditional roosting grounds. Animal Alliance of Canada and other animal welfare societies denounced the plan as cruel and convinced the mayor to stay the execution and investigate alternatives. The city council wanted more control of a situation that was putting their town at odds with animal lovers across the nation. The council took matters into its own hands, relieving the mayor of an apparently self-appointed duty. Some residents grew impatient, most conspicuously the mayor's younger brother, Richard Erickson. He decided to organize a rural crow shoot, complete with a hundred-dollar prize to the hunter who bagged the largest crow. The hunt was legal. Crow season is always open, and in this part of Canada there are no limits on how many crows can be shot. The Chatham City Council voted fourteen to two to give the public works department authority to move the urban crow roost by whatever means were needed. They tried the large-eyed scarecrow kites from Japan. They hazed the roost with blank shotgun shots. They killed a few crows and hung them around favorite crow haunts. Nothing worked but Erickson's "crow derby." Hunters killed a few crows, much to the environmental community's disgust. The public's conflicting views about crows were coming face to face. The inability to identify a crow correctly was also apparent, since the winner of the derby bagged a bird so big, weighing more than two pounds (one kilogram), that it was likely a raven. Based on a tip, the police raided Erickson's house, seized his gun, ammo, and two marijuana cigarettes, and eventually charged him for possessing a loaded firearm in a vehicle and discharging a firearm within eight yards (meters) of a roadway. Crow shooting died out and the council began to debate other options.

The council hired a professional wildlife specialist in 2001. A falconer working for Bird Control International succeeded where others could not. Using owls, hawks, and eagles, he disturbed the Chatham crow roost to the point that only 147 birds remained during the winter of 2001–2002.[38]

The Chatham Crow War shows humanity at its best and worse against nature. Patience and persistence paid off as the environment that once favored crows was changed. The safety of the roost, originally provided by regulation against shooting in towns, was reversed by the introduction of trained predators. By understanding the mechanism responsible for this wildlife-human conflict, an efficient, quick, and relatively inexpensive solution was found. Brute force alone rarely works to deter crows. Stopping access to superabundant foods, or increasing the risk of living with us can be effective.

EIGHT

Centering the Balance

On September 6, 2002, our morning's field work found American Crows flourishing across a four-year-old neighborhood we had been studying twenty miles east of Seattle, along the city's rapidly sprawling "urban fringe." We had watched this neighborhood since it was a two-hundred-acre (eighty-hectare) patch of forest rarely penetrated by crows. Now it was a subdivision supporting sixty-eight new households and regular flocks of cruising, foraging, nonbreeding crows. By the summer of the following year crows would be breeding in the new backyards.

That afternoon, we received a fax from the U.S. Fish and Wildlife Service in Hawaii that underscored how the balance of crows is shifting dramatically on our rapidly peopled planet. Biologists Jeff Burgett and Dave Ledig advised us that searches over three days for the last pair of Hawaiian

253

Crows ('Alala) flying free in the forests above Kona had come up empty. The pair had not been seen the previous two months. The 'Alala was definitely extinct in the wild. The only other two species of crows ever to live on Hawaii never survived the Polynesians' occupation of Oahu, Maui, and Molokai, and now the last of the most powerful forest voices there was silent. The Hawaiian ecosystem at Kona would depend on avicultural science, luck, and fifty captive 'Alala for the restoration of this eloquent corvid to the moist forests of the Big Island.[1]

Most crows and ravens adjust to our actions. People and crows reinforce one another's actions, often for mutual gain, as the birds infiltrate our cultures and ecologies. We spill garbage, and Jungle and American Crows can expand their geographic ranges, cultures of movement, and diet. Their populations grow, and they gain momentary leads in their evolutionary races. Other interactions are less mutually reinforcing. Americans' vast gardens of corn, wheat, and soybeans transformed the Great Plains, forcing Common Ravens to live in remote areas where sloppy predators still provided scraps for scavengers. But even such massive reconstruction of a continent's land cover has neither extinguished the raven nor caused it to evolve detectable genetic changes. Which raises intriguing questions: If the fate of corvids is so closely tied to humans, might we be driving the way they evolve as a family, pruning some species from the corvid tree and encouraging branching, or speciation, elsewhere? And if so, can we find evidence of evolutionary change in their genetic makeup? This is no mere scientific parlor game. For as humans alter the habitats, ecosystems, and the rise and fall of the creatures in them, we may be forcing large-scale genetic changes in the species we come into contact with. Corvids, a family of creatures so closely aligned with our culture, offer a unique window into the ability of humans to alter, however inadvertently, the genetic makeup of the world around us. Such

With the American Crow shown at top for comparison, current and former crows of Hawaii include the 'Alala, the extinct Deep-billed Crow, and the extinct Slender-billed Crow

changes would constitute evolutionary connections between people and corvids that are rooted in *genetics* rather than culture.

This chapter takes up the hunt for genetic responses of corvids to people. We illustrate these responses where they naturally happen most quickly, on islands. Islands profoundly stimulated the evolutionary theories of Charles Darwin and Alfred Russel Wallace. They also illustrate our most extreme current effects on the evolution of wild crows. When people colonize islands, we forever change their plant and animal communities. We shuffle the cards of the evolutionary deck in extreme ways by introducing new species and extinguishing old ones. In little over a thousand years we have extinguished more than half of all the bird species that occupied the lush islands of the tropical Pacific. This includes the extinction of a population of 'Alala that formerly lived on Maui, the impressively large Deep-billed Crow (*Corvus impluviatus*), with a monstrous, arched bill from Oahu, and the Slender-billed Crow (*Corvus viriosus*) from Oahu and Molokai. Yet the casual visitor, inundated by a bewildering diversity of birds, scarcely notices. Look carefully and you will see that these are exotic birds introduced by people, not native birds created by island evolution. The extinction and introduction of birds, mammals, and plants on islands create new ecosystems and stimulate new evolutionary interactions. Island crows are caught in these new interactions. Often they are among the last native species to remain. Their shrunken populations may speed innovation or extinction. Either way, evolution is rapid, and we are ultimately responsible for changing the genetic composition of populations, species, and whole families of birds.[2]

Genetic evolution occurs when a population's genetic makeup changes in response to novel challenges from the environment. Populations with sufficient genetic variation may have a few individuals that are able to exploit or adapt to novel changes. Such individuals pass their fortunate genetic combinations on to their offspring, which outnumber the few, if any, offspring produced by less adapted individuals. Through time, by survival *and reproduction* of the fittest, the population shifts to include a higher proportion of the genetic combinations held by the original, fortunate few. This is the process of natural selection.

Through the vast majority of time, nature provided the novel challenges that organisms adapted to. Currently, humans have taken over as the instigators of challenge. Our unprecedented mobility, in combination with our carelessness and ecological naïveté, rapidly challenges less mobile species. Consider the seemingly benign dumping of excess water by sailors at tropical ports of call. This was common during the nineteenth century. By this simple action, Mexican sailors released mosquitoes in a stream on Maui in 1826. Alone, mosquitoes pose little threat to a species as large as the 20-ounce (650-gram) Hawaiian Crow, but as a vector for other pathogens like malaria and avian poxvirus, themselves brought to Hawaii in the early twentieth century in non-native birds, they provide an insurmountable challenge. Malaria and pox are thought to be primarily responsible for many recent extinctions of Hawaiian birds. Most native species survive best above four thousand feet (thirteen hundred meters), where cold may limit the prevalence of disease and abundance of mosquitoes. However, even this limit is slowly creeping upward as mosquitoes adapt and global temperatures rise.[3]

Hawaiian Crows were fortunate to live at moderate elevations on the Big Island (Hawaii), where malaria and pox were slow to find them. But humans began to persecute them and change their habitat in the twentieth century. Together with disease, they reduced the range of 'Alala from the moist, midelevation, southwestern half of the island to three small, moist forests, 3,200–6,500 feet (1,000–2,000 meters) above the sea on Mauna Loa's western and southern coasts. About a hundred 'Alala could be found in these forests in the early 1970s. Only eleven existed in 1993 when Marzluff first saw them on the McCandless Ranch and lawsuits forced herculean recovery efforts. These efforts have not yet restored 'Alala to the wild, hence Burgett and Ledig's fax of September 6, 2002.

'Alala were unable to meet the challenges people presented to them.

'Alala are tropical forest specialists that eat an extensive array of fruits, nuts, insects, and the eggs and nestlings of other birds. This diet, however, is not what doomed them. Rather, malaria weakened the population and increased the crows' vulnerability to predation by rats, exotic mongoose, and even the native Hawaiian Hawk, or 'Io (*Buteo solitarius*). People introduced rats, which eat crow eggs and young, and then added mongooses to eat the bothersome rats. Rats are eaten by Hawaiian Hawks, which also thrive in

Natural predators like the Hawaiian Hawk ('Io) have contributed to the extinction of the Hawaiian Crow ('Alala) in the wild

the easily hunted forest openings people make. Mongooses eat everything that they can get their sharp teeth on, including the contents of crow nests and recently fledged, but barely mobile young crows. Farmers shot considerable numbers of 'Alala, and a parade of settlers during the last thousand years burned, cleared, logged, and grazed the forest, reducing its suitability as crow habitat. Feral house cats from homes and farms also brought the toxoplasmosis parasite to the crows, which severely limited restoration attempts. Together these factors extinguished the 'Alala from wild Hawaii. Human challenges were too novel, too diverse, and too rapid for this species to adapt.[4]

In the process, people affected the 'Alala's genetic evolution simply by stopping it in nature and starting it in captivity. Since September 2002, all existing 'Alala have lived in captive breeding facilities. Their evolution is now completely guided by the "ecosystem" we have made for them and the challenges contained therein.

In the most extreme way possible, we now guide the evolution of the 'Alala. We select which crow breeds with which mate and which individuals will eventually repopulate the wild. In the future, this species "tinkering" might well include selecting for hardy crows that have resistance to malaria. We are molding the gene pool of 'Alala to keep it as diverse as possible, hoping that genetic diversity will increase the crows' chance of solving future natural challenges. We participate in and support these efforts even at an annual price tag exceeding two million dollars. Without captive breeding and eventual reintroduction, the 'Alala will forever be lost and people will have profoundly affected the evolution of the corvid lineage.[5]

The Hawaiian Islands have taught us much about the devastation that often follows careless introduction of exotic species. But if you really want to experience an island ecosystem silenced by humanity's actions you must fly west another 3,600 miles (5,900 kilometers). As you leave the plane's coolness and enter Won Pat International Airport on the island of Guam, you first notice the heat and humidity. But you don't notice the Jack Russell terriers sniffing cargo to find snakes or the barriers around the airport that are the first line of defense in new efforts to contain accidental introductions of exotic species. Guam, an American territory in the southern Mariana Islands, is on full alert because in the late 1940s or early 1950s, as the World War II ended and a jubilant American military moved supplies from Melanesia to Guam for storage, they also moved a thin, nondescript, nocturnal snake.

The introduction of the Brown Tree Snake (*Boiga irregularis*) to Guam has all but silenced the dawn chorus of birdsong. The snake reached unprecedented abundance on Guam—up to 13,000 snakes per acre (32,500 per hectare)—where it has no natural predators. A voracious predator on birds and their eggs, this snake ate twelve species of Guam birds to extinction during the late 1970s and 1980s. Seventeen of eighteen native species were severely affected by snakes. Only the Mariana Crow, known by local Chamorro people as *Aga*, Mariana Swiftlet (*Aerodramus bartschi*), Micronesian Starling (*Aplonis opaca*), Brown Noddy (*Anous stolidus*), White Tern (*Gygis alba*), Common Moorhen (*Ballinula chloropus*), and Yellow Bittern (*Ixobrychus sinensis*) survived, and only barely. Aga are small by crow standards. At about nine ounces (250 grams), they weigh little more than a jay. But their coal-black plumage, large bill, and hoarse voice betray their crow roots. As with the 'Alala, the Aga is a forest crow that eats a variety of plant and animal matter, including insects, crabs, fruits, and nuts. An "abundant"

The introduced Brown Tree Snake has decimated Guam's birds, such as the Mariana Crow, whose nest is shown here

population of Aga, perhaps twenty-five hundred birds, in the mid-1940s was reduced to around a hundred in the 1980s, twenty-six in 1995, and ten in 2003. Without experimental translocations of crows from the nearby island of Rota, where around five hundred birds existed in 2003, the species would likely be extinct on Guam.[6]

Even restoration of the Aga on Guam will change the course of evolution. To our knowledge, Aga existed only on Guam and Rota. The Rota population was probably founded by a recent colonization event from Guam. Since that time, genetic evidence suggests that these two populations, separated by thirty miles (forty-nine kilometers) of deep Pacific Ocean, have remained isolated and distinct. Although both island environments are similar, slight differences and chance events produced some detectable genetic variations between the Aga populations on Guam and Rota. Recent molecular analyses revealed that Aga on Guam were more diverse than those on Rota—a surprising result because there were many fewer Aga on Guam than on Rota. This lack of genetic diversity on Rota may reflect an earlier evolutionary response of Aga to people. Fighting and bombing during World War II obliterated much of Rota and probably killed most Aga. Crows may even have been exterminated from Rota at this time. Whether a few dispersers from Guam or a few survivors on Rota reestablished the Aga on this small island is uncertain. But either sort of "bottleneck" in the population size of Aga would lower genetic diversity. The recent repopulation of Guam with Aga from Rota is again causing evolutionary change in the Aga by replacing a relatively diverse Guam gene pool with a less diverse Rota gene pool. To be sure, conservation biologists, ourselves included, reason that replacing an important ecological link like the Aga, even if it causes some genetic rearrangement of the species, is warranted. In fact, restoring the Aga may be doubly important because not only are we replacing an important ecological force, we likely are replacing an important cultural force for island people.[7]

Still, human effects on the evolution of island corvids are profound because the birds' numbers are naturally small and therefore susceptible to even minor challenges from their environment. It is true that small populations can evolve quickly because a few successful individuals with high reproductive rates can cause large changes in a small gene pool. This sort of successful exploitation of new "challenges" has produced much of the splendid diversity for which islands are famous. However, as the Aga and the 'Alala have shown us, small populations can also rapidly decline to the brink of extinction. Other less well known island crows like the Palm Crow, White-necked Crow, and Cuban Crow from the Caribbean archipelago may also be facing extinction. By creating impossible challenges for small and specialized populations we are profoundly affecting the genetic composition of island crows and shaping the evolutionary tree of corvid life through careless and indifferent pruning.

Influencing evolution on islands is not the only way we shape crows. We also affect the relative composition of species on large continents. Earlier, we alluded to the possible role of humans in the evolution of the Northwestern Crow. This intertidal crow that resides from northwest Washington to southeast Alaska cannot be reliably distinguished from American Crows by voice, behavior, or morphology. As recently as a hundred years ago Northwestern Crows were isolated from American Crows by impenetrable lowland forests. Presumably they were evolving along a unique path as they solved the challenges of living at the northern hemisphere's only interface between a cold, productive ocean and a temperate rainforest. This novel environment apparently favored, among other things, small body size and colonial breeding. The Northwestern Crow's unique evolutionary path is now also being traveled by the American Crow. Our settlements and agricultural transformation of the middle and western portions of North America allowed American Crows to spread west and inhabit towns and stream corridors in inland Washington at the turn of the twentieth century. Forest clearing by intensive logging, increased settlement, and a mobile human population facilitated the movement of American Crows to Washington's coasts a few decades later. Interbreeding is now likely to be regular between these two "species," but by definition, true species do not regularly and successfully interbreed, which means that any unique aspects of Northwestern Crows are quickly being lost as abundant and successful American Crow traits flood the gene pool. This hypothesis awaits molecular confirmation, but to biologists like ourselves who live in the Northwest, it is obvious that our region's evolutionary balance has shifted away from unique, colonial, intertidal crows toward larger, more mobile crows that thrive amid human settlements, campgrounds, and dumpsters while routinely dining along the coast's rocky shores.[8]

Our last example of genetic responses of crows to people is the most speculative, but also the most indicative of future interactions. It started late in the summer of 1999 in New York City. Apparently a person, pet bird, or living mosquito traveled from Israel to New York and brought to North America a new, exotic disease. Known as West Nile virus, this native of the Congo quickly spread across North America killing millions of birds, hundreds of people, and thousands of pets and zoological specimens in its first four years. West Nile virus is a pathogen that infects animals, including people, after they are bitten by an infected mosquito. This virulent strain of virus is especially deadly to birds. The West Nile virus has infected more than 160 species of wild birds, but none has been as devastated as the American Crow.[9]

Crows contract West Nile virus from infected mosquitoes, infected hipboscid flies that live naturally in bird's feathers, infected prey, and even from each other. Virtually every infected crow dies. Studies of marked wild crows in New York, Oklahoma, and Illinois documented annual mortality rates of 30 to 70 percent after West Nile virus first appeared. That's six to twelve times the low mortality rate typically enjoyed by the species. In recent years, tens of thousands of dead crows have been collected by an alert public, but even this level of mortality is barely detectable at larger geographic scales. Some regional crow population counts in the Northeast and Midwest

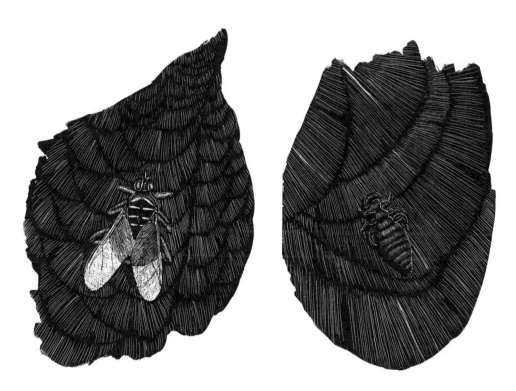

Crows preen themselves vigorously to remove parasites like this (a) hipboscid fly and (b) a feather louse evolved in color to match the crow's plumage. These external parasites are possible vectors of disease, like West Nile virus

have declined steadily from 1999 to 2002, but individuals remain throughout the species' range, local breeding populations are relatively stable, and young crows continue to be fledged, even in areas hit hardest by the virus. Thus, American Crow population explosions may be dampened by our introduction of an exotic virus, but it is unlikely that American Crows will go extinct.[10]

Large and widespread continental species rarely go extinct because the entire species usually doesn't meet a new environmental challenge all at once. Many American Crows have not met West Nile virus, despite its rapid romp across the United States and Canada. They may meet it next year, but many crow families or local populations may never meet it because the virus has a patchy distribution determined in part by the occurrence of mosquitoes and the occurrence of mosquito control activities. Crows in dry areas or areas where people aggressively control mosquito larvae may have a selective advantage over crows with more mosquito contact. Even where breeding crows are killed, the abundance of crows and the short season of abundant, infected mosquitoes means that some crows will survive and swiftly fill in open breeding space. The social system of crows allows those who die to be replaced quickly by nonbreeding helpers or floaters, or perhaps even by individuals migrating from distant, less affected areas.

The response of crows to West Nile virus may be subtle genetic and cultural change. Recent studies of mice documented genetically based resistance to West Nile virus. Perhaps the crows that survive exposure to West Nile virus are genetically distinct from those that succumb to the high fever and swelling of the brain, encephalitis, that kills others in a slow and pathetic death. Genetic traits that allow some individuals to survive are the stuff of evolution. We have no evidence yet to support or refute this idea. Nor do we know that West Nile virus will change crow culture, but we suspect it will. Those crows that avoid large communal roosts near mosquito-infested wetlands or those that preen their partners' hippoboscid flies less often may be less susceptible to the disease. These sorts of behavioral change may spread through a population by simple natural selection. Unlike cultural evolution, a blind leader rather than a knowledgeable teacher directs these behavioral changes. They may even have a genetic basis and therefore eventually lead to genetic changes in crows after continuous exposure to West Nile virus.[11]

It took us only one night to see the possible evolutionary force of West Nile virus and other diseases. We watched a sick crow die under a large roost in Brier, Washington. Other sick birds were evident in the lower reaches of the cottonwoods and willows that grow in the swampy area. We suspected that some of those near the dead and dying were healthy mates or offspring. At least twenty-five crows had died at this spot during the past week, and their smell reminded us of rotting salmon along favorite spawning streams. As night fell, about five hundred crows clustered in a few large trees. We watched them with binoculars and saw them surrounded by clouds of mosquitoes that had risen from ground level to envelop the roosting birds. A few minutes later, with darkness almost complete, we were surprised to see two crows, silhouetted against the sky, as they bolted from the roost and flew into the night. Their antisocial behavior may have been a blessing in disguise. By roosting away from mosquitoes and dying flock mates, perhaps these crows would survive. Whatever incited such behavior might then be favored. Communal roosting, such a defining characteristic of American Crows and a behavior that has served them so well in the past, might die out in this age of rapidly spreading exotic diseases. As it happened, these crows appear to have been dying from a reovirus (respiratory enteric orphan virus), which is known to affect many animals in a variety of benign to deadly ways. But just as reovirus might shape the evolution of the Brier crows, next year it might be West Nile virus. Either way, in Brier's urban setting it seems likely that people are ultimately responsible. The myriad possibilities, including viruses, pollution, poison, and pesticides, underscore the likelihood —near certainty—of our effects on the genetic evolution of crows.

The endangerment of island crows and recent expansions of American Crows have produced a variety of responses by people concerned with restoring our increasingly disrupted ecosystems. These are the latest in a long history of cultural responses by people to crows. Usually they reverse our previous transgressions. In this way, Common Ravens are being returned to parts of Europe where only a century ago human persecution removed them. Our study of 'Alala, Aga, and American Crows shows how they are being restored or limited. It also suggests several simple ways to keep the full complement of our native crows and ravens moderately abundant, rather than overly abundant or uncomfortably rare. Sustained change will be more difficult until people adopt biocentric values, those that promote the survival of all species not just people. But we are optimistic that the powerful influence crows have on our culture can help people value other species. Those values will be needed to maintain a diverse assemblage of plants and animals on Earth, generally, and a diversity of native crow species, specifically.[12]

Limiting expanding populations of American Crows is challenging. To have even moderate, lasting success we need to do more than just shoot crows and bomb roosts. These actions are unacceptable to a growing portion of our society, and they are insufficient to limit a large, mobile, and adaptable population that often associates with our children and pets and frequents our parks. Our long-standing bad habits need to be changed if we are truly interested in limiting expanding species like American Crows so that species like Fish Crows and Northwestern Crows will have room to evolve and rare species like Marbled Murrelets will have space to breed. Although this tall order will probably never be fully attained, ideas we have mentioned earlier in the book suggest simple strategies to limit our enabling of the more exploitative corvids. There's something for everybody in taking action for the collective corvid good.[13]

Reducing a crow's access to our garbage, compost, and bird feeders will certainly affect the numbers of breeding and wintering crows in our neighborhoods. This can be easily done by locking lids on your trash, using a covered container for composting food waste, and using only relatively enclosed bird feeders, stocked principally with seeds and grains. Crows prefer bread, pet food, or table scraps, so keep those out of your feeders. Corvids often steal pet food if it is easily available, so feed your pets small meals that are readily consumed, and cover any extra food.

This may seem like a killjoy suggestion for people who see feeding birds as a way to interact with nature. So, how do we reconcile these competing notions? Dare we suggest that part of the fabulous cultural coevolution between people and crows be actively avoided? Yes, in situations where the crows that are being fed are such rapidly expanding species as American Crows, House Crows, and Jungle Crows. By helping to increase the populations of such species, we may actually be lowering the overall diversity of crow species and increasing the negative interactions between other people and crows. These unintended consequences can erode public support for environmental conservation, distance people from nature, and broadly reduce the positive cultural coevolution often seen between people and crows. We suggest that if you feed crows, at least do it responsibly. Know which species you are feeding and understand if there are potential negative implications of your actions on other species or other people's values. If you are feeding an expanding species of crow, then do it sporadically in a restricted area. Occasionally supplementing a few pairs of crows is less likely to fuel population growth than is consistent feeding of many crows.

Private citizens, landscape architects, urban planners, and developers can limit crows simply by reducing lawn cover and increasing shrub and tree cover in yards, parks, and neighborhoods. Think of the extra time and money you could save if you had only a small lawn to mow. Landscaping with native shrubs increases habitat for small songbirds and reduces crow habitat. It also saves you the hassles and ecological insults of applying herbicide, pesticide, and excess water on plantings. Our studies suggest that small lawns, less than about a quarter of an acre (a thousand square meters), surrounded by trees and shrubs, though still used by robins, towhees, juncos, sparrows, and small children, are rarely used by crows, especially if homeowners do not provide other foraging opportunities as discussed above.

City officials, business owners, park managers, and recreationists hold an important key to limiting expanding crow populations. Reducing the availability of food in urban and suburban areas could dramatically increase overwinter mortality of crows, especially naive juveniles. Requiring restaurants to dump their refuse in well-sealed hard containers would be prudent. Landfill managers could reduce accessibility to abundant food by covering garbage frequently and stringing wires in meshlike grids above open pits. Wires are effective at dissuading crows, ravens, and gulls that quickly learn to avoid harm by staying away from the confining mesh. At a smaller scale, park managers can increase the use of animal-proof garbage cans and educate visitors about the negative ecological implications of feeding wildlife. While outdoors, be extra cautious to keep your food away from even the most solicitous and appealing wildlife. As much fun as it is to have a jay take a cracker from your hand or a crow sidle up to your sandwich, it is not worth the potential damage to other less tolerant wildlife. Our studies with Erik Neatherlin, then a University of Washington graduate student, suggest that crows first get their beak into remote wilderness locations by colonizing campgrounds and exploiting the riches brought to them by campers and picnickers. Others have even suggested that junk-food addictions by seed-dispersing corvids, like Clark's Nutcrackers, may disrupt their important ecological role as tree-dispersal agents. Nutcrackers content with available campground fare may harvest and store few seeds, some of which would have gone unretrieved and germinated to contribute to forest expansion and recovery.[14]

Our biggest challenge is limiting crows in and around agricultural areas. The uncanny ability of these birds to use portions of our crops is virtually unstoppable. Even extreme measures are ineffective. The 3.8 million crows killed by dynamiting 127 roosts from 1934 to 1945 in Oklahoma yielded no detectable reduction in crow populations or agricultural damage. Less drastic measures, such as discouraging with noise, scarecrows, distress calls, predators, and dead crows, do not keep crows from expanses of exposed grains for long. Our only recourse is to reduce the availability of grain, which is usually not cost-effective. Most farmers simply acknowledge a certain loss to crows and other birds each year. Fortunately, some loss is repaid by the crow's reduction in harmful insects, as discussed earlier. A moderate amount of "predation" on intensively domesticated crops like corn might even be beneficial. For instance, by selecting for diversity in agricultural practices and mixtures of seed sources, vulnerability of our crops to pathogens and unpredictable weather may be less than if we plant only monocultures. A single crop planted in vast acreage can easily be devastated by a predator adapted to consume it or by slight, but intolerable changes in weather.[15]

The most feasible way we can limit the crow use of crops is to limit crows in our settlements and recreational sites. Our garbage fuels crows of all ages, all across the country, at all times of the year. This stimulates population growth by increasing reproduction and survival. It may also buffer and enhance populations outside settlements. Living in rural lands is harder for crows; they have to fly further and are killed more often by people and predators. Forcing most crows to live in rural landscapes may stabilize their populations at current numbers, rather than increasing them. Historical increases in rural lands occurred as we increased agriculture. Today, agricultural lands are shrinking in many regions nationwide, so keeping crows rural may keep them in check. Even moderate-sized rural crow populations will continue to challenge our farmers and limit their yields, but this is a price we owe nature. We think the cultural stimulus and ecological services these cunning birds provide enrich our lives and keep our crops healthy and diverse.

When we slip from moderation to rarity, the real costs of poorly stewarding our wildlife resources become painfully apparent. Such is the case on the Pacific Islands where 'Alala and Aga are being restored. Restoration of island crows involves five basic steps. We have to identify and eventually remove factors that limit the growth of crow populations. We must also manage wild populations to increase their productivity. Young crows must be reared in captivity, and captive adult crows must be housed and co-erced to breed. We may then release independent birds back into the wild. Last, we monitor crow reproduction and survivorship and measure population growth. Each step involves a small army of dedicated scientists, aviculturists, managers, policy makers, and landowners. As mandated by the U.S. Endangered Species Act, a federally appointed recovery team coordinates actions. Aga and 'Alala recovery teams include ten to fifteen knowledgeable and concerned scientists, federal and state agency and nongovernmental organization representatives, and private landowners. These teams advise the U.S. Fish and Wildlife Service, the agency in charge of endangered species recovery on lands in the United States and its protectorates, on how best to restore crow populations to self-sustaining, free-living status. To claim that all agree on how best to restore island crows would be far from the truth. But all have embraced a fairly straightforward course of action that combines intense management and research on wild and captive populations with slow, often politically charged attempts to neutralize known limiting factors.

A curator at the San Diego Zoo once told us that raising crows is easy; in fact, she insisted, you can't even "kill them with a shovel." Our experience proved to be a quite different story. Once you have a fertile crow egg, especially if a parent crow has provided the first week of incubation, you almost always hatch a vigorous chick. These chicks grow nicely if you carefully feed them with a mush of blended native fruits, boiled eggs, insects, and baby mice every thirty to sixty minutes, twelve to fifteen hours a day, seven days a week, for their entire thirty-to-fifty-day-long nestling period. However, stunting, crippling, and occasionally death result if the food is not carefully balanced for vitamin, mineral, and water content, or if a small amount accidentally gets into a begging chick's windpipe, or if common yeasts, bacteria, or fungi attack small nestlings, or if the rapidly growing chicks are not sunned, exercised, or housed in properly lined and sized nest bowls. If you can avoid these pitfalls, then hatching and raising baby crows is pretty easy. It just takes a dedicated staff, time, and money.

Puppets are used during captive rearing of endangered Hawaiian Crows

The hard parts are getting viable eggs and releasing young birds that are savvy enough to survive in the wild and breed. Viable eggs are obtained from carefully observed wild pairs or from carefully housed captive pairs. Each source provides a challenge. Consider the wild crows. First, the eggs need to be fertile. The last survivors of a species are often old and physiologically past their prime. Remaining wild Aga and 'Alala pairs routinely laid infertile eggs, making efforts to hatch them futile. Second, the eggs need to be moved from the wild to the lab. This means they must survive about a week in predator-laden nature, be handed down from a nest ten to twenty yards (meters) up in a tree, and transported many miles (kilometers) over inhospitable terrain to a waiting incubator. It is not uncommon to lose a nest before eggs are pulled, but once an egg has reached a biologist's hands, it has always made it safely to the incubator. The eggs are picked out of the nest by the biologist wearing surgical gloves, packed in a padded case, and lowered to ground crews. They are then packed in portable incubators or warmed millet seed and carried to a truck, plane, or helicopter, where they are transported to the rearing facility several hours away. Field crews feel pressure mount with each bounce of the vehicle. Marzluff's longest ride down Mauna Loa was when he and Peter Harrity, a biologist for the Peregrine Fund, drove the last 'Alala egg of the 1993 season to the incubator.

Stealing eggs from the last nests of a wild population may seem ludicrous, but we are careful to make sure that this does not jeopardize the safety or future reproduction of the wild pair. Common Ravens and American Crows have taught us that removing eggs early in the nesting season actually stimulates renesting by the robbed pair, much as naturally happens when a predator or typhoon destroys a nest. So, egg removal itself is an important tool used to increase the productivity of declining populations. The first clutch goes to the lab for rearing while subsequent clutches remain with parents, thus doubling annual egg output.

Captive breeding and rearing programs for Aga and 'Alala have been successful. From 1993 to 2000, thirty-seven young 'Alala were produced, which more than doubled the world's population. Likewise, eight adult Aga, twelve eggs, and thirteen nestlings were moved from Rota to Guam from 1996 to 2002. Three out of every four Aga moved from Rota has survived long enough to hatch, grow, and be released on Guam. Without these efforts Aga would be gone from Guam today.[16]

Notwithstanding these successes, getting released Aga and 'Alala to restore once vibrant wild populations remains an elusive goal. We learned some of the nuances about releasing island crows by experimenting with American Crows, Common Ravens, and Black-billed Magpies in the Idaho shrublands. Marzluff and his students Kate Whitmore and Laura Landon parented hundreds of corvids for three years. We raised and released them after various periods of care, in groups of various size, and in various locations. We exposed birds to a variety of foods and predators before release. We learned that an effective strategy was to release groups of three to five birds, a few months after they could forage for themselves, from field aviaries, where they acclimated for a month or more before release. Birds released in this way weaned themselves from our food, water, and shelter at their own pace and were usually successful at avoiding the numerous natural and human obstacles common to the rural Idaho landscape. Obstacles included hawks, eagles, owls, coyotes, water troughs, electricity, and rifles. Survivors integrated with nearby wild birds. A few that we were able to keep close track of bred early in their lives and were successful parents. But Idaho is not Hawaii—released island crows have not been able to avoid the threats or circumvent the changes within their environments. Released Aga are beginning to pair, and a few have laid eggs in the wild, but released 'Alala have succumbed to disease (toxoplasmosis), exotic species (mongoose), and the abundant native hawk. While these releases help us understand the many factors that must be reduced to restore natural populations, increasingly they demonstrate that beyond purely biotic effects, successful recovery of endangered species requires effective leadership, determination, and public cooperation.[17]

Simply put, island crow restoration has not yet succeeded because it has not captured the imagination of enough people. Brown Tree Snakes, rats, mongooses, cats, and diseases still kill crows in the wild. Controlling these agents of destruction, though technologically challenging, is achievable with enough financial and political support. Current island recovery efforts are grossly underfunded relative to efforts to restore species on the American mainland. From 1992 to 1995, for example, the U.S. Fish and Wildlife Service spent around one hundred thousand dollars annually on recovery of each endangered island species but more than two million dollars on each more widely distributed, less threatened, mainland species. It's harder to save an island crow than a continental wolf, Peregrine Falcon, or Bald Eagle. Politicians are unable or unwilling to effectively campaign for island restoration in general and island crows specifically. We suspect any campaigning that occurs falls on deaf national ears because islands have little clout in Washington, D.C. Of all the Pacific and Caribbean Islands, only Hawaii has formal representation in Congress, but because few people live in Hawaii, its political influence is minimal. As a result, presidents and their appointed Fish and Wildlife Service directors gain little politically from successful island restoration projects and, accordingly, invest little in them.[18]

Greater federal investment in island recovery efforts is necessary, but not sufficient, to save what we think of as "paradise." Island residents will have to make difficult choices if they are to keep their native fauna. Much of their land, the most precious of island commodities, will need to be managed to sustain populations of native plants and animals. Some human use of land will need to be sacrificed. Sacrifice can have some lasting advantages. Recently, landowners and state agencies have benefited from increased ecotourism, profitable land sales to conservation interests, and cooperative land restoration that produces a healthier and more attractive ecosystem. Such benefits should be celebrated and expanded with additional funding, incentives, and creative thinking. With more incentives to practice restoration on private lands, more landowners and managers will likely embrace it. There will be some of course who are rigidly fixed to a consumptive mode and cannot entertain the conservation of anything, much less an endangered crow. If they prevail, all of our lives will be impoverished by the growing loss of biological diversity that such an attitude fosters.

The sorts of active restoration and limitation we have discussed are necessary to maintain diverse ecosystems in today's human-dominated world. As a society, however, we need to go further. We need to make cultural responses to the loss of wild Hawaiian Crows, decline of Mariana, Cuban, Palm, and White-necked Crows, and genetic swamping of Northwestern Crows. Specifically, we need to value these species as integral components of the ecosystems upon which we all depend. Crows in particular are powerful reflectors of ecosystem function. They alert us to devastating and costly exotic species, act as sentinels to dangerous pathogens, and by using them to their own advantage quickly point out our most wasteful habits. Their presence early in our cultural formation remains a powerful influence on our lives. Our language, art, religion, and pop culture would be different, and we think poorer, without crows. Today society often focuses on the negative influences of crows like their costs to our agricultural production, their noisy and irritating habits, or the sacrifices we must make to restore rare island species. This single-minded focus fosters disdain of the species and increasingly disconnects people from nature. Our reduction of the diversity and distribution of crow species may therefore actually stifle creative aspects of our own cultural evolution. Keeping a wide variety of crow species moderately abundant may reduce our negative interactions with a powerful cultural motivator and allow us to continue our positive and mutually reinforcing cultural coevolution. Maintaining positive and creative associations with crows may do more than continue to stimulate our culture. It may enable us to appreciate more of nature's splendor. Appreciation of nature strengthens our natural predisposition toward *biophilia,* literally the love of life, which

Hawaiian Crow ('Alala). Drawing based on a photograph from 2003 by Jack Jefrey.

is a prerequisite for sustained conservation of our ecological support systems. The first Hawaiians kept 'Alala in their villages and considered them guardian spirits, or *'aumakuas*. That crows should remain for future generations to consider such possibilities seems fitting for people still seeking to reconcile themselves with nature. Henry David Thoreau grasped this in 1869 when he noted that the crow "sees the white man come and the Indian withdraw, but it withdraws not. Its untamed voice is still heard above the tinkling of the forge. It sees a race pass away, but it passes not away. It remains to remind us of aboriginal nature."[19]

NINE

Future Interactions

Our discussions about people and crows, those handsome, forty-some species of the genus *Corvus,* have led us to a unique overview. We see a relationship between people and animals that is multifaceted and runs both ways. We occasionally protect our crops and health from crows, but our interactions go much beyond mere practicality. Our experiences with crows, for instance, are far more diverse than our strongly utilitarian relationships with fish, whales, and livestock. At a basic level, human society is a powerful engine forcing ecological conditions and evolutionary pressures on crows. We increase the survivorship and reproduction of many crow species by transforming Earth's land cover. We speed the extinction of others by exposing them to disease, predators, and competitors, including other crows. Such responses of crows to people become especially interest-

Ravens assemble around early hunters, anticipating an opportunity to scavenge

ing when we consider the additional cultural connections existing between these birds and ourselves.

These strong, long-standing ecological and evolutionary linkages between crows and ourselves have fostered a unique pattern of coevolution between us. Other species that affect some human cultures, like whales, bears, wolves, cranes, and eagles, struggle on with their lives in spite of us. Many are endangered by our actions today. Crows usually live with us, and not just with some of us. Crows, because of their ubiquitous association with people, affect most human cultures. Their conspicuousness, curiosity, sociality, insightful behavior, and powerful voices have captured our imaginations more significantly than other species. Reviewing our discussions about crows and people therefore conjures up an image of a diverse array of people interacting with a host of crow species through time and across the world. Most interactions are of people affecting the ecology, culture, and evolution of crow populations. Effects of crows on the culture of people are also obvious, but effects of crows on the ecology and genetic evolution of people are more tenuous. This is not unexpected. Our sociality, learning and teaching abilities, and ability to modify Earth for our needs means that today people change culturally much more frequently than they change genetically. Therefore crows really affect us where it counts—in our culture. And we affect crows where it counts for most nonhuman species—in their ecology, culture, and even genetic composition.

As we introduced you to the world of crows, we pointed out many examples of how we think our cultures have coevolved. We suggest that six are particularly well documented and best illustrate our idea of cultural coevolution. We review these below as a summary of our ideas and as a way to think about how our cultures may continue to coevolve in the future. We offer these case studies as working hypotheses. We do not doubt the cultural connections and changes we describe, but in many cases we cannot be absolutely certain that a specific human culture stimulated a specific behavioral response in a crow population that is truly evolving by cultural transmission. Distinguishing social learning from independent trial-and-error learning is especially difficult. We anticipate that our premise of cultural coevolution between people and wild animals will give momentum to the search for further evidence of its existence. Most certainly we would hope for a careful consideration of the examples we provide.

1. Hunting and gathering. Preindustrial people were intimately connected to nature. Nature was their world and crows and ravens were integral players. Intersection between the human culture of hunting and gathering wild food and the corvid culture of scavenging from predators was simply unavoidable. We were just another producer for corvids to exploit. Cultural coevolution resulted as people began to use the behavior of crows and ravens to inform them about their surroundings. Many people must have followed ravens to kills, but some, like the Eskimos of Greenland, were said to have been led by ravens to good hunting opportunities. If such guiding and following occurred, it would be an example of close cultural coevolution where the behavior of ravens changed to facilitate following by people and the behavior of people changed to understand and respond to the raven's signals. People and ravens both would have benefited from increased foraging efficiency. Mutual benefit would have fueled increasingly close association. Such coevolution has occurred between people and birds in Africa, where the Boran people and the Greater Honeyguide mutually facilitate each other's abilities to exploit the resources of wild honeybees, albeit from individual rather than social learning by honeyguides.[1]

As prehistoric peoples strengthened their relationship with crows and ravens, our culture celebrated the birds as gods, creators, and tricksters. But they also came increasingly to be viewed as competitors for valuable foods. Human culture responded to the corvid culture of raiding food processing and caching locations by erecting some form of scarecrow and chasing offending birds. These tactics have changed over time, but they were present at least a thousand years ago in Mimbres settlements, and a hundred years ago in Makah villages of the northwest Washington coast. Early responses to crows were controlled, for taboos often prevented killing crows and ravens.[2]

A scarecrow adorns a Makah salmon-drying rack in Neah Bay, Washington. Based on a photograph from 1900.

Keeping food away from raiding crows and ravens may have helped form an even more basic aspect of human culture, our social lifestyle. The advantages of using children to shoo ravens from stored food and the need to process a hunter's kill quickly before scavengers took a substantial portion certainly would have favored cooperative hunting and gathering, group living, and unique jobs. Crows and ravens, in part, could have literally made us what we are today. Whether early humans scrounged leftovers from African carnivores, confronted and overpowered predators at their kills, or defended scavenged meat in home bases, they would have confronted corvids. Pied Crows (*Corvus albus*), Fan-tailed Ravens (*Corvus rhipidurus*), and White-necked Ravens (*Corvus albicollis*) lived on the savannas, riparian woodlands, mountains, and deserts of Africa where humans evolved from two to seven million years ago. Perhaps humans cooperated to keep their precious bone marrow and meat meals away from pesky corvids. The large mammals of Africa would have posed serious threats to evolving humans, but even rudimentary cooperation would have been effective against crows and ravens. A start to the cooperative lifestyle in response to small competitors like corvids may have prepared early humans to compete more effectively with deadly and formidable mammalian predators and scavengers. No fossil corvids have yet been found at early human archaeological sites. Delicate bones and lack of teeth make bird fossils generally rare. But we anticipate their discovery when environments that preserve a wide variety of animals are discovered. Indeed, fossil ravens are found with saber-toothed cats in later, North American sites like Rancho La Brea.[3]

Important interactions between humans and corvids are more certain in the late Pleistocene (500,000 to 250,000 years ago). At that time, European Neanderthals and ancient *Homo sapiens* were actively hunting antelope and other large mammals throughout Europe and Asia. For evolving humans, this was a time of considerable and rapid brain growth, probably in response to the demands of social life and big-game hunting. Wolves and ravens would have been common and persistent competitors. We think it is likely that the raven culture of persistence, thievery, and caching would have coevolved with meat transport, storage, and defense strategies of humans.

Eventually, humans incorporated their corvid associates into legend, cave art, myths, and ornamentation.[4]

Our far-reaching speculation about the roles of corvids in early human evolution comes from an investigation of Gray Wolf (*Canis lupus*) sociality. Wolves are almost always in the company of scavenging ravens. Ravens are even called "wolf birds." Wolves are highly social, cooperative, and communicative. But pairs of wolves, not the larger and more familiar packs, have been shown to be optimal for hunting. A wolf maximizes its energetic return from hunting by sharing its prey with only one other wolf. Packs of unrelated, cooperating wolves are not strictly required for successful hunting, and they require kills to be shared. Packs have many benefits, one of which may be defending their kills from Common Ravens. Ravens quickly gather after wolves make a kill, and begin to eat and cache available meat. Depending on the number of ravens, groups of wolves routinely lose five to forty-five pounds (two to twenty kilograms) of food per day to ravens. As wolf pack size increases, however, losses to ravens decrease. These observations led researchers in Michigan to conclude that the costs of sharing food among wolves in a large pack are more than offset by the increased food retained from ravens. Wolves and ravens are clearly culturally coevolved. Pleistocene humans and wolves faced ravens together, first in Europe and then in North America. As ravens molded hominid sociality, perhaps they also favored a more intimate cultural coevolution—domestication. The wolf, our first domestic species, may have joined ancient families, in part, to chase away ravens.[5]

The modern melding of hunter-gatherer societies with industrial societies appears to have eroded some cultural connections to nature. Loss of shamans and traditions have weakened spiritual connections among some native people and crows and ravens. As our culture changes, so does our ecological relationship with these birds. Crows that raid fish stores of the Quinault Nation today are routinely killed. Ravens that raid Koyukon villages are also occasionally killed. But killed ravens are apologized to or scolded and forgiven for such lowly behavior. Cultural reverence seems to be giving way to apathy or animosity. We suspect that this increasing disconnection of people from nature will continue to foster a culture of disrespect for crows and ravens. Corvid culture will likely respond with cautious avoidance of threatening human settings and association with favorable settings, as we discuss later in the case of urbanization.[6]

 2. Expansion of agriculture. The human culture of clearing the lands of standing forests and sowing them with an abundance of domesticated plants ignited a cultural revolution in crows. Crows quickly accepted these crops as a new food source and changed migratory patterns to favor their consumption. Much of the culture of today's American Crow is a direct response to our ancestors' agrarian culture. Like the ravens that adapted to hunter-gatherer culture, crows increasingly came in contact with and tolerant of agricultural society and eventually began to seek it out. This cultural change

Plowing the prairies provided a bridge to the crow's domination of North America

was met with resistance by some who built scarecrows, or hunted and harassed offending crows. Other people accepted corvids and incorporated them into religion, art, and pop culture. A coat of arms from medieval England, with a raven perched on top of a sheaf of wheat, bears the title "Live Eternally." Allowing crows and ravens such a prominent place speaks to the reverence some received even in an agrarian culture. Crows adjusted to resistance by shifting to locally unprotected crops, hiding and carefully defending their nests, modifying roost approaches, and recognizing new and subtle dangers—like men with guns versus men with rakes. In some cases, crows abandoned rural settings for less hostile urban settings.[7]

3. War and aggression. Throughout history, wars have left battlegrounds littered with corpses, thereby providing the opportunistic scavenger with a superabundant food source. The more carnivorous corvids, like ravens, magpies, and Carrion Crows, were ready and waiting. Our culture of warfare nicely fit the foraging culture of these corvids. Seeing crows scavenging the dead, people began to draw literary, spiritual, and artistic connections between corvids and death. For many, crows and ravens now became evil harbingers of death, with connections to the supernatural. To some degree corvids were cast in the role of scapegoat for the suffering and degradation that accompanied war. Increasing hostility toward crows ensued, and the birds responded by retreating from settlements and shying away from people. This pattern was especially evident in Europe during the nineteenth and twentieth centuries. Persecution so reduced the abundance of many corvids that a human culture of restoration has now developed. People are working to bring ravens back to regions of northern Europe and the southeastern United States. Rarity has also produced a culture of tolerance or ambivalence toward many once-shunned corvids that are now returning to urban areas across the continent. Rarity may also drive people to long for

Crows, ravens, and vultures readily feast on corpses

what was once both familiar and mysterious and so name favorite lodges, getaways, and neighborhoods after crows and ravens. Our culture may always associate corvids with death as we watch them scavenge a variety of dead animals, but our association of corvids with human death will likely fade, or live only in pop culture, as increasingly sanitized wars are fought and our corpses are quickly buried.[8]

4. Urbanization and recreation. Crows and ravens have shared our settlements for thousands of years. Carrion Crow, Common Raven, and human remains are commingled in ancient settlements (4,000 to 10,500 years ago) at

Troy, in Mesopotamia, and in modern-day Syria, Poland, and western Canada. We expect corvid-human cohabitation to strengthen as humans become increasingly urban. By 2030, more than 60 percent of Earth's population is expected to live in cities. This new urban-intensive culture of people has several effects on crows and ravens. Our conversion of forests, grasslands, and coastal areas to burgeoning settlements and scattered recreational sites provided new foods that corvids culturally incorporated into their diets and foraging habits. These include earthworms, exotic fruits and seeds, refuse, and pet food. Our buildings, utility poles, and other structures provided new nesting and roosting sites. Our sewage ponds, watered lawns, and lakes bring life-giving resources to these birds in arid regions. Such subsidies allow many corvids to exploit new environments or attain enlarged populations where they were formerly rare. For example, Common Ravens are now nesting on oil-well towers and shelter structures on the North Slope of arctic Alaska, on buildings in the cities of Phoenix and Los Angeles, and on utility poles, billboards, and bridges in the Mojave Desert and Great Basin shrublands. Use of anthropogenic nesting sites, especially in regions where cliff and tree nests are otherwise rare, has allowed raven populations to increase exponentially, expand into new regions, and reclaim parts of their formerly extensive geographic range that were relinquished to crows a century earlier. Intrusion of human activities into Earth's remaining remote locations will exacerbate the raven's expansion. The roosting culture of many corvids, including Common Ravens, Rooks, Western Jackdaws, Hooded/Carrion Crows, and American Crows has also responded to the warmth, protection, and vertical structure that our cities provide. A winter evening in the Potsdamer Platz district of Berlin is highlighted by the arrival of thousands of Rooks, Western Jackdaws, and Hooded Crows that descend onto the glass skyscrapers and street trees to roost communally in a warm, safe location.[9]

Refuse becomes a rich resource as cities grow. We tend to concentrate our waste in a variety of ways, thereby facilitating its exploitation by corvids. Urban corvids have adopted a culture of scavenging our refuse, including an uncanny ability to quickly recognize many novel food types and learn how to get at garbage held in cans, dumpsters, bags, and boxes. Corvid social culture also responds to our trash. Most corvids evolved to exploit a natural variety of seasonal foods, but our refuse is abundant, concentrated, and reliable. Corvids responded to this source of food with natural adjustments in their typically territorial behavior. Dominant corvids do not defend abundant and reliable foods; defense is too costly and ineffective. Species like American Crows, Common Ravens, and Jungle Crows therefore form large, relatively placid flocks at our dumps. Flocking behavior of non-breeding corvids may be especially responsive to reliable garbage. Young ravens, for example, form more stable flocks at refuse dumps than at less reliable and smaller animal kills. Dependable and abundant food may also lead to delayed dispersal and increased family size in American Crows.[10]

Roads and automobiles are a defining feature of large cities. Our culture of driving and paving has been exploited by corvids who habitually forage along our roadways for dead and injured animals. The culture of ravens in remote areas with roads is noticeably affected. Each morning adults fly directly above roads in their territories looking for the preceding evening's carnage. Their offspring likely learn this habit by following them on road patrol soon after they can fly. Our cars have replaced many of the large predators that scavenging corvids evolved with, but with slight cultural adjustments, these new providers are also exploited. In cities, cars become even more valuable to some crows. Automobiles function as insect nets and nutcrackers. Bold American Crows pick bugs off car grills in parking lots, and savvy Carrion Crows in Japan carefully place thick-shelled walnuts in front of stopped cars, where they are effectively crushed as soon as traffic moves. The cultural roots of car use by Carrion Crows is being investigated, but the fact that the behavior has spread several miles (kilometers) along the Hirose River in three decades from its origin at a driving school in 1975 is consistent with cultural evolution.[11]

People have responded to the cultural adjustments made by corvids to city life in basic and now familiar ways. We harass and kill corvids when populations exceed our "cultural carrying capacities" in settlements and at recreational sites. We attempt to reduce corvid use of refuse and buildings by inventing "animal proof" waste cans, putting wires above dumps, enclosing our garbage in bags and nets, covering and incinerating our waste,

Urban crows quickly adapt to new sources of food. Here they pick dead insects off a car's front grill

and installing spikes, covers, and electric wires to reduce use of perch and nest sites by a host of birds, including crows and ravens. Occasionally, we catch and remove offending birds. One Hooded Crow from Dortmund, Germany, that attacked people was tricked into eating alcohol-laced cat food. The drunk crow, who fared better than Hans Huckebein's unlucky raven, was captured and sent to an animal shelter to sleep it off. Sometimes we allow corvids to use our structures but modify the structure to reduce the damage caused by birds. In this way, many miles (kilometers) of power lines in southern Idaho have sported shields that keep raven feces from coating the lines and short-circuiting the system. Some of us form personal relationships with these intriguing and dominant city birds. In Japan, for example, some people aim their car tires at the walnuts those Carrion Crows place in the road, thereby reinforcing the culture of nut-cracking in a culturally coevolutionary way. Finally, our popular and scientific culture responds to the increasing familiarity of our corvine coinhabitors. Corvids star in movies, serve as mascots for our urban sports teams, inspire music groups, and even encourage some street people. Their ability to live with us piques our scientific curiosity to learn how they exploit us when so many other species cannot.[12]

Corvid culture responds to our adjustments by circumventing most of our control efforts and gradually adjusting their distribution to areas where people tolerate or encourage them. They get through or under barriers, change roost and nest locations to areas without persecution, and adjust the intensity and conspicuousness of nest defense to our response. Because ambivalence and encouragement of corvids is greater in urban than rural areas, we expect corvids to parallel our increasingly urban ways well into the future. More mobility for people may even actively facilitate corvid dispersal among our urban centers in the future. People often long for trappings from their homelands. As greater numbers of corvids become familiar and interesting partners for urban people, we suspect that some will be moved between cities and released into the wild. This has happened in the past as British colonists moved Rooks from England to New Zealand. More crows in one location, however, may reduce the overall diversity of the species as aggressive species outcompete others.[13]

    5. Hunting for recreation. The culture of crow hunting, passed through our society in spoken stories, books and Web sites, has produced a series of cultural responses by crows. Hunted crow populations exercise considerably more caution around people than unhunted populations. Hunted crows recognize hunters, their vehicles, and guns. They modify their roost approach and perhaps roost locations to minimize mortality. In response to an increasingly wary crow culture, crow hunters try new calls, decoys, guns, and shot loads. We asked a local hunter how he bagged one of our banded crows and learned the extreme form this arms race can take. This hunter devised a special lure using crow and pigeon wings and tails fastened to a black board that he placed below an owl decoy. While hiding in a nearby barn, he broadcast recordings of crows mobbing owls until a group of crows swarmed around the suspected predatory owl. Using a silent air rifle, he killed seven crows before the confused group dispersed. Crows likely will become increasingly concentrated in areas that

A proud hunter bags a crow

prohibit hunting and harvest if these activities accelerate elsewhere in the future. Given a resurgence in popularity of crows as traditional dishes in places like Lithuania, cultural coevolution with hunters may alter the behavior of a wide range of crow species in the future.[14]

6. Bird feeding. The culture of feeding wild birds is practiced by a rapidly growing segment of society. Corvids adjust their behavior swiftly to exploit novel food offered at a variety of feeders. Anyone with a feeder can attest to the fact that crows culturally pass the behavior of using feeders and the food contained therein to their offspring, which noisily accompany parents to feeders each summer shortly after fledging. People respond strongly to crows at the backyard bird feeder ("bird table" to Europeans); many attempt to dissuade the gluttonous birds by hazing or installing counterbalanced perches that close their feeder when large birds or mammals, like squirrels, attempt to eat. One acquaintance keeps a single dead crow in his freezer. By displaying it prominently for a few days whenever a crow visits his yard, he keeps most crows at bay. Other backyard birders encourage crows by providing special food and feeders. Occasionally this sort of cultural coevolution leads to personal relationships between people and crows.

Limiting a crow's time at the bird feeder helps keep their numbers in check

In the United States, for instance, it seems that each neighborhood has at least one resident who makes a daily ritual of feeding crows specially delivered nuts, pet food, or table scraps. Crows solicit food from these people by congregating at their residence and calling. People respond by whistling or yelling encouragement to the crows to come ever closer. Often crow families will interact with such people for years or decades, eventually nesting nearby and even engaging pets and family members in games. We suspect that some friends, neighbors, and family members of crow-feeding people also take up the habit, which then spreads via social learning.

The culture of feeding crows need not spread easily through the human population. Neighbors, employers, and even legal threats may pressure people who feed crows to stop. In 2003, a panicked school bus driver from Bellevue, Washington, called us for advice on how to keep her job. For twelve years Phyllis Alverdes fed American Crows on her way to work each day, attracting a noisy, black horde. Crows had developed a culture of following her car and begging for handouts—a culture that Alverdes's employer did not appreciate. Fearing that the messy birds would damage his busses, he gave her an ultimatum: stop feeding crows or lose your job. Other bus drivers, the local transportation workers' union, and concerned citizens rallied on Alverdes's behalf. Her job was spared, but her car was no longer permitted in the bus lot. Alverdes's effect on the culture of crows reverberated through human culture as people refined their opinions about the value of wild animals and determined the reasonable rights of workers. The duality of cultural responses to clever but mischievous animals like crows is unlikely to wane. Some people will always hate crows, while others will love them. Crows will flee from the former and gravitate to the latter.[15]

Taken as a whole, the behavioral response of corvids to people approaches what even the most ardent humanist would require of a "culture."

The rich variety of crow behaviors, their persistence and evolution through the ages, and the likely importance of individual innovation and subsequent social learning bestow on crows a heritage of knowledge, customs, capabilities, and habits that allows them to keep pace with humans. We have focused on individual crow behaviors in our examples of corvid culture, stressing how they respond to and reciprocally challenge human culture. But the whole of these many cultural elements may well be greater than the sum of the individual parts.[16]

Cultural coevolution may eventually confer genetic distinction on corvids. Behavior often changes first to solve novel environmental challenges, but if selection remains strong and movement of individuals is not excessive, then local customs may become genetic legacies. In this way, the sorts of cultural coevolution we have discussed can create new corvid species. Perhaps future people will know new, genetically distinct species like the "Urban Crow," "Clam Cracker Crow," "Road Crow," "Cemetery Raven," or "Maize Rook." With sufficient genetic isolation to limit interbreeding, speciation seems possible for two reasons. First, globalization increases the homogeneity of our culture, thereby offering extensive corvid populations the same selective pressures. Second, these selective pressures are increasingly severe and distinct between rural and urban settings. Our urban culture, as well as the diseases we disperse, predators we introduce, and climate we change, can send urban crows on evolutionary trajectories quite distinct from their rural counterparts. Our environmental challenges to nature generally, and crows specifically, shape populations that can adjust. And extinguish those that cannot. Cultural coevolution is part of the creative aspect of our interaction with crows; extinction of small, specialized species, like those often found on islands, is the destructive counterpart.

The balance between human-forced creation and extinction will determine the diversity of crows in the future.

Our interaction with crows and their kin is an ongoing saga, and it's rather tantalizing to fantasize about what the future may bring. Pet stores of tomorrow may offer you a choice of parrots, ducks, chickens, and crows. Some of us may be recycling the nutrients and energy from our garbage with tame ravens that consume it. Trained corvids could fly out to spot, circle, and flush game for hunters. Raucous "watch crows" might alert us to the presence of a daytime intruder. However appealing these options may be to some people, we doubt that the demand will be great enough to domesticate the likes of crows and ravens. In the past we have domesticated animals that are of manageable demeanor and high utilitarian value. Corvids, unlike chickens, turkeys, pigeons, ducks, and geese, just do not offer us enough utility to balance their high maintenance costs. To many they are not as beautiful or eloquent as a parrot. To most they are not as plump and tasty as a chicken. Americans will find it against the Migratory Bird Treaty Act to possess one. Crows certainly will never be as obedient as the dog or as cuddly as a cat. This is not to say that a crow raised singly from the time it is a young nestling does not make a good pet. It does. It follows its adopted human like a puppy, runs to it for protection, may sleep in the human's or other pet's bed, affectionately preens hair expecting reciprocity, and assumes that it should act as humanlike as possible. But even this appeal will never override a quintessential corvid trait destined to keep crows and ravens on the edge, rather than in the center, of our lives.[17]

That defining trait is the corvid's inquisitive nature, which often leads to mischief. It seems that tame corvids cannot resist prying, probing, pulling, pushing, or penetrating anything that has the likelihood of opening up or having something beneath its surface. Past generations warned of this inquisitive mischief by calling crows and ravens "tricksters." Current crow fanciers are quick to report on the havoc their pets wreak. We know of many pet ravens that routinely strip the rubber from any available windshield wiper in the owner's neighborhood. Those allowed free exit from the home are infamous for pounding on the neighbor's house at the crack of dawn, stealing laundry, or tormenting the nearby dogs and cats. Bill Gilbert wrote about his pet crow, Hello, which stole cigarettes and spectacles, finished off fuzzy navels (orange juice and peach schnapps), and cached spoons, spark plugs, coins, pencils, eyeglasses, rings, and beads. Pete Byers's crow, Edgar Allan Crow, showed an interest in ichthyology. Edgar "collected" the family angelfish and stored them by pressing them neatly between the pages of an old book for safekeeping, just as botanists do with plant specimens. It is our inability to take the mischief out of the trickster that we think will keep corvids wild into the future.[18]

Keeping corvids wild keeps our relationship with them interesting. It may also keep our relationship with nature, in general, healthy. As we have argued, people and crows share a remarkable ecological, evolutionary, and cultural history. These fascinating birds may have played a large part in sustaining our connection to nature and all the social, psychological, and physical benefits we are just beginning to document and appreciate. We are learning that people who can see or reach forest from their home tolerate denser housing developments, hospital patients exposed to the sights and sounds of nature speed to recovery, and aging men and women stay healthier if they interact with animals. Corvids living among us are the best of nature's ambassadors as the web of our ancestral connections to nature becomes increasingly frayed. Their complex and mysterious voices, insightful problem-solving behaviors, and curious prominence provide an intriguing and convenient means to link urban people with wild nature.[19]

Raven profile

Perhaps the gods gave us crows to provoke and remind us how foolish it is to assume an all-knowing human ascendancy over nature. Indeed, there's no doubt that our curious descendents will, like ourselves, be lured to the wood and streams within and beyond the city by these birds to, as Thoreau said, "confront only the essentials of life." The birds will remain to soothe our urban souls, stimulate new artisans and dreamers, and, we hope, provide future generations of people with the wisdom to maintain healthy ecosystems on their home planet.[20]

In spite of people's efforts to study them, the crow and raven remain a collection of perplexing contradictions. To some people they are just those noisy trash birds that steal our crops, spread garbage in our streets, eat our favorite songbirds, and annoyingly wake us up too early. To others they are mystic messengers who can warn us of danger, carry souls of friends to the afterlife, and stimulate our art, language, and pop culture. Still others see them as emissaries of a deity, scrutinizing our actions, trying to guide our future, reporting on our progress, and shaping our destiny. To many they fulfill each of these roles. Indeed, confronting the physical and spiritual power of these amazing birds, while defying explanation, will provide fire for the imagination. Our wonder over crows and ravens has allowed them to figure prominently in our culture. After living for sixteen months with the Koyukon people of interior Alaska, Richard Nelson summed it up beautifully: "What is the raven? Bird-watchers and biologists know. Koyukon elders and their children who listen know. But those like me, who have heard and accepted them both, are left to watch and wonder." Our wish for you is that you keep watching and wondering. As long as people watch and wonder about the natural world, our culture will continue to be enriched by the antics, mysteries, and challenges of crows and ravens in all their forms.[21]

Appendix 1

Making Observations to Learn More

Unless you live in South America or Antarctica, you probably live with some crow or raven species. You and your family's day-to-day interactions with these birds can help us all better understand our unique relationship with crows. So we invite you to challenge, probe, and test our assertions about crows and people by engaging in careful study of these amazing birds. We would love to know what you find out (mail to: Marzluff, Box 352100, University of Washington, Seattle, WA 98195) but suggest that you first look for interested collaborators in local bird clubs, like the National Audubon Society (http://www.audubon.org), museums, colleges, and universities. You might learn about opportunities to join national or local counts (e.g., http://birds.cornell.edu), find another kindred soul in need of assistance, or partner with someone just eager to learn more about the ways of crows. Below are some suggestions to guide your curiosity.

Before you start, plan how you will document your findings. A journal of observations is an important central repository for your thoughts and observations. Note the date, time, and weather conditions associated with your observations. Be systematic. For example, if you regularly walk a set route, note the abundance of the birds, or the occurrence of specific behaviors, along the route each day, month, or year. Bird counts are best done at the same time of day and for a standard length of time. Put your observations on a map so you can determine where local crow abundance waxes and wanes. Look for environmental and human factors associated with crow occurrence. Negative information, the failure to find a species in a particular area or observe it not doing a particular behavior, is very important. Noting where corvids are not or what they do not do is just as important as noting what they actually do and where they do it. (Lugingbuhl et al. 2001 detail a survey protocol specifically for counting corvids. Using this protocol would allow your results to be compared easily with those done by other researchers.)

Much still remains to be learned about the basic natural history of crows and ravens. In 1943 Joseph Hickey, a former professor at the University of Wisconsin, published the engaging book *A Guide to Birdwatching*, which gives a detailed set of questions to guide any study of bird natural history. We suggest that you follow Hickey's advice and document regional occurrences of crow species and note what sorts of habitats each species inhabits. Describe the habitat, noting vegetation, physical landscapes, and human attributes. Search for crows and ravens in various haunts. Compare changes in their use of areas as the areas themselves undergo transition. When development or recreation increases, do crows respond? Note how much area you cover and how long you spend searching for a species in a particular spot. Take note of when crows or ravens first colonize a new area, such as crows in desert areas and ravens nesting in remote arctic regions or busy cities. In this way you could develop standardized counts of crows and document changes in abundance over time. Follow birds to their evening roosts to estimate population size and document the local environments used for roosting. Ask nearby residents what they think of the crows. Note mortality sources and cooperate with local game or health agencies, respecting their protocols for handling the dead birds, to document the occurrence and spread of new diseases that affect crows, such as West Nile virus. Document the variety of foods corvids eat, perhaps focusing on their role as predators and nest raiders. By watching a sample of other birds' nests you could gather important information on the frequency of crow predation. Does the rate of nest predation relate to qualities of the preyed-on nests, such as their visibility or proximity to homes? Do you detect changes in the numbers or nest attempts of the birds being preyed on?

Perhaps the most interesting sorts of observations you can make involve novel behaviors used by wild crows. The antics of your pet crows can also be insightful, but often they are harder to interpret because of possible cuing by their owners. In addition to detailed notes about the behaviors you observe, we encourage you to document your observations with video or still photography. A picture or video is invaluable if accompanied by detailed descriptions about when, where, and how it was obtained. New digital technologies allow you to easily record and submit images to interested scientists. Point your cameras at vocalizing crows and ravens to catalog their rich vocal repertoires and associated postures. Try to connect a specific vocalization with a setting, context, and possible stimulus. If done in a variety of places, you might notice regional dialects within a species. If you can discriminate call types, then document, by counting during a fixed time period, their relative occurrence with respect to area, season, individual, and context. Record instances of apparent play in your local corvids and film their reactions to friends and foes. Do they recognize individual people or other animals in their environments? See if you can document some of the weird social gatherings we hear about, like funerals, visits to people, and executions.

Investigate the learning and problem-solving abilities of your local birds. Document the specific sorts of fast-food bags they recognize. Record any use of tools. Tell how they gain access to your garbage or bird feeder. Because it is so difficult to document actual cultural transmission of behaviors, pay special attention to situations where birds learn to perform adaptive tasks by watching or following each other rather than learning by trial and error. We hypothesize that this frequently occurs, but you can help test the hypothesis. Be on the lookout for behavioral differences among crow populations.

Whenever it is feasible, try to replicate your observations on many individual crows or in many locations so that you can appraise the generality of your results. Also, compare your observations to the behavior of crows during "control" periods. For example, if you find a dead crow and wish to investigate funeral behavior, record the behavior of birds before, during, and after you present the dead bird to them and contrast this change in behavior with their response to another object like a can or a dead animal other than a crow that you present in the same way and monitor for the same length of time.

Conservation would be served if you could use your knowledge to suggest ways to limit expanding crow species from exploiting our resources. Perhaps you could test a variety of scarecrows or show us how to keep crows out of our bird feeders and crops. Use your insight and knowledge to educate your neighbors about living in crow country.

Of course, we would be most interested in learning about the myriad ways crows and ravens interact with people in your part of the world. Record the sorts of human resources your crows use. Knowing more about historic cultural connections between people and crows would be fascinating. Report your local stories, legends, and myths about crows. Catalog any ancient renditions of crows on nearby caves or interesting effects of crows on your local popular culture.

Many of the ideas we suggest would be suitable for publication or make good science fair or school projects for kids. If you are interested in publishing your most interesting findings, you first need to learn what others have published. The References section of the book contains many scholarly readings to introduce you to the primary ornithological literature. Once you know how your observations fit with published accounts, we suggest that you select a relevant journal and contact its editor. There are many international, national, and regional bird journals. Each has a specific focus and writing style that the editor (or often the journal's Web site) can tell you about. One of the most important things you can do is to encourage children to explore crows so that this next generation forges a tighter bond to nature. There are many children's books to kindle the interests of budding natural and cultural historians. We list some of our favorites below and wish you good reading.

Appendix 2

Children's Books That Involve Crows and Ravens

Armstrong, J. 1995. *King crow.* New York: Crown.

Blassingame, W. 1979. *Wonders of crows.* New York: Dodd Mead Wonder Books.

Boyd, L. 1998. *Lulu Crow's garden: A silly old story with brand new pictures.* Boston: Little, Brown.

Cameron, A. 1987. *Raven returns the water.* Madeira Park, BC: Harbour.

———. 1991. *Raven and Snipe.* Madeira Park, BC: Harbour.

———. 1991. *Raven goes berrypicking.* Madeira Park, BC: Harbour.

Carlstrom, N. 1997. *Raven and river.* Boston: Little, Brown.

Chorao, K. S. 2000. *Pig and crow.* New York: Henry Holt.

Clark, P. N. 2003. *In the shadow of the mammoth.* West Bay Shore, NY: Blue Marlin.

Cunningham, D. 1996. *A crow's journey*. Morton Grove, IL: Albert Whitman.

De Felice, C. 1998. *Clever crow*. New York: Simon and Schuster.

DeLage, I. 1983. *The old witch and the crows*. Champaign, IL: Garrard.

Dillon, J. 1992. *Jeb Scarecrow's pumpkin patch*. New York: Houghton Mifflin.

Dixon, A. 1992. *How Raven brought light to people*. Basingstoke, UK: MacMillan.

Erdrich, L. 1999. *The birchbark house*. New York: Scholastic.

Farmer, N. 2004. *The sea of trolls*. New York: Atheneum.

Frascino, E. 1988. *Nanny Noony and the magic spell*. Toronto: Pippin.

Freeman, D. 1960. *Cyrano the crow*. East Rutherford, NJ: Viking.

Gadd, B. 2001. *Raven's end*. Toronto: McClelland and Stewart.

Gage, W. 1984. *The crow and Mrs. Gaddy*. New York: Greenwillow.

George, J. C. 1980. *The cry of the crow*. New York: Harper and Row.

————. 2004. *Charlie's raven*. New York: Dutton.

Grimm, J. 1994. *The seven ravens*. Orlando, FL: Harcourt Brace.

Guy, G. 1991. *Black crow, black crow*. New York: Greenwillow.

Hale, I. 1992. *The naughty crow*. New York: Margaret K. McElderry Books.

Harsh, F. 1991. *Alfie*. New York: Ideals.

Hayes, S. 1992. *Crumbling castle*. Cambridge, MA: Candlewick.

Hazelton, E. B. 1969. *Sammy, the crow who remembered*. New York: Charles Scribner's Sons. [Note: Sammy is actually a Raven!]

Hobbs, V. 1999. *Carolina crow girl*. New York: Farrar, Straus and Giroux.

Holder, H. 1992. *Carmine the crow crows*. New York: Farrar, Straus and Giroux.

Hyman, R. 1989. *Casper and the rainbow bird*. Princeton, NJ: Barron's.

Jacobs, L. 2003. *Crow*. San Diego, CA: Blackbird.

Latimer, J. 1992. *James Bear's pie*. New York: Charles Scribner and Sons.

Lopez, B. 1990. *Crow and weasel*. San Francisco: North Point.

Luenn, Nancy. 1993. *Song for the ancient forest*. Basingstoke, UK: Macmillan Atheneum.

Marion, J. D. 1992. *Hello, crow*. Longfield, UK: Orchard.

Martini, C. 2004. *The mob*. Tonawanda, NY: Kids Can Press.

McDermott, G. 1993. *Raven: A trickster tale from the Pacific Northwest*. Orlando, FL: Harcourt Brace.

———. 1994. *Coyote: A trickster tale from the American Southwest*. Orlando, FL: Harcourt Brace.

McDonald, M. A. 2000. *Jays*. Broomall, PA: Child's World.

Melzack, R. 1970. *Raven, creator of the world*. Boston: Little, Brown.

Miller, E. 1977. *Raven helps the Indians: A Skokomish legend*. Beaverton, OR: Educational Systems.

Orgel, D. 1995. *Two crows counting*. New York: Bantam.

Pinkney, A. D. 1998. *Raven in a dove house*. Orlando, FL: Harcourt Brace.

Pringle, L. 2002. *Crows! Strange and wonderful*. Honesdale, PA: Boyds Mills.

Robinson, Gail. 1982. *Raven the Trickster: Legends of the North American Indians*. New York: Atheneum.

Rosen, M. 1995. *Crow and hawk: A traditional Pueblo Indian story*. Orlando, FL: Harcourt Brace.

Rowe, J. A. 1994. *A baby crow*. New York: North-South Books.

Savage, C. 1995. *Bird brains*. Vancouver, BC: Greystone.

Schami, R. 1996. *The crow who stood on his beak*. New York: North-South Books.

Scuderi, L. 1998. *To fly*. Brooklyn, NY: Cranky Nell Books.

Spalding, A. 2000. *The keeper and the crows*. Victoria, BC: Orca.

Van Laan, N. 1989. *Rainbow crow: A Lenape tale*. New York: Knopf.

Wakefield, A. 1996. *Those calculating crows!* New York: Simon and Schuster.

Weigelt, U. 2001. *It wasn't me!* New York: North-South Books.

Woodruff, E. 2003. *The Ravenmaster's secret: Escape from the Tower of London.* New York: Scholastic.

Yashima, T. 1955. *Crow boy.* Middlesex, UK: Penguin Books.

Notes

ONE

Cultural Connections

1. Marr and Calisher (2003) put forth the interesting argument that Alexander the Great died in 323 BC from West Nile virus. This disease is especially deadly to corvids, which might explain the flock of dead ravens at the Babylonian gate. In addition, Marr and Calisher report that Alexander's chills, abdominal pain, and partial paralysis, in addition to the mosquito-infested Babylonian swamps, are consistent with infection by West Nile virus. Others question Marr and Calisher's diagnosis, claiming that typhoid is more likely (Dewar 2003), but the dead ravens are an interesting clue pointing to West Nile virus.

2. Crow-headed man is from Armstrong (1958), who discussed European mythology and crows in detail. Aborigines and ravens are discussed by Goodwin (1978).

3. Osgood (1971) details the Hän; Freuchen and Salomonsen (1958) describe the Greenlandic Eskimo; Sax (2003) tells of Tibetan funeral riturals.

4. Armstrong (1958) reviews European mythology; Sax (2003) discusses representa-

tion of families and clans with crow and raven figures and discusses derivation of "corone."

5. Clarke 1998–2002, http://www.r-clarke.org.uk/constellations/corvus.htm, shows the constellation *Corvus;* Wells (2002) relates Hebrew views of crows; Lee (2000) mentions Crowbone in his discussion of Norwegian history; and Martin (1993) discusses Eastern and French views of specific crow sightings. Cassidy (1984) details crow augury with examples of what timing, number, and direction mean.

6. Shirota (1989) discusses Japanese crow roosts.

7. Tribe of Crow is a story reported in *Real Change* (*Tribe of Crow* 1999). Quotation by Twain (1897). Kurosawa et al. (2000, 2001, 2002) detail crow outbreaks in Tokyo.

8. Rick Knight's work is detailed in Knight (1984) and Knight et al. (1987).

9. Culture and social learning in animals are discussed by Bonner (1980), Boyd and Richerson (1985), Rendell and Whitehead (2001a), and Shennan (2002). Fritz and Kotrschal (1999) present a good example of social learning in captive ravens. Our definition of culture is from Boyd and Richerson (1985) and Rendell and Whitehead (2001b). The importance of public information to social learning is presented by Danchin et al. (2004). For a good discussion of those supportive and those critical of culture in animals other than primates, see the responses to Rendell and Whitehead (2001a) that occurred in the journal *Behavioral and Brain Sciences* 24 (2001): 324–382. Perspectives on human cultural evolution, the benefits of cultural versus genetic evolution, and how genes and cultures become coevolved are reviewed by Durham (1991), Feldman et al. (1996), Odling-Smee et al. (2003), and Castro and Toro (2004). Durham (1991) details the interplay between culture and genes in the evolution of human lactose tolerance.

10. Cultural transmission is discussed by Dawkins (1976), Blackmore (1999), and Mesoudi et al. (2004). The relative composition of memes, reflecting genetic, individually learned, and socially transmitted information, is discussed by Durham (1991) and Laland et al. (2000). We view memes as units of information transferred among individuals by social learning (e.g., rules, songs, religions, and even such specific behaviors as handshakes and dietary choices [Dawkins 1976; Blackmore 1999]). Although Blackmore insists that memes must be imitated exactly to evolve culturally, we do not agree. Traditions can evolve by inexact social learning, including stimulus enhancement and local enhancement where only the general behavior or location is learned (Danchin et al. 2004). Archaeology and culture are discussed by Shennan (2002).

11. Deeke et al. (2000, 2002); Barrett-Leonard et al. (1996).

12. Imo's story can be found in Avital and Jablonka (2000). Many examples of whale and dolphin culture are recounted by Rendell and Whitehead (2001a).

13. We recognize that all behavioral traits of animals are not cultural, but because few airtight demonstrations of cultural evolution now exist in crows, we are admittedly loose in our categorization of behavior as culture. Cultural designation is rarely absolute because genes may quickly coevolve with memes, and individual and social learning occur simultaneously in long-lived social species (Lachlan and Feldman 2003; Odling-Smee et al. 2003). Individual trial-and-error learners become the models for social learning. Therefore, in a society where cultural inheritance is at work, some, but not all, individuals may acquire behavior by social learning. Without detailed, usually long-term, research that documents acquisition of behavior by recognizable individuals, the occurrence of social learning is difficult to prove without a shadow of a doubt. However, in social species where individuals capable of sophisticated learning regularly aggregate and cooperate, some social learning seems inevitable. In fact, social learning is usually the most parsimonious explanation for the acquisition of behaviors that develop in social settings (e.g., foraging, roosting, and mobbing behaviors). Accordingly, we attribute many corvid behaviors to social learning and regularly invoke cultural evolution as a process. Goodwin (1978) discusses the culture of bread eating.

14. New Caledonian Crows are researched by Hunt (1996, 2000a,b), Hunt et al. (2001), and Hunt and Gray (2002). Nut-cracking and shell-dropping by corvids is from Zach (1979) and Cristol and Switzer (1999). Kenward et al. (2005) demonstrate that hand-reared, naive New Caledonian Crows use stick tools. This may be a case of gene-culture coevolution where the basic propensity to use tools has evolved a strong genetic bias (Lachlan and Feldman 2003), but the particular choice of tool type and technique of use remains flexible to local culture.

15. Crow song dialects are discussed by Brown (1985) and Brown and Farabaugh (1997). Lorenz (1952) reports his thoughts on recognition of people by corvids.

16. Altshuler and Clark (2003) describe a hummingbird-plant coevolution.

17. Our view of cultural coevolution extends gene-culture coevolutionary theory (Durham 1991; Laland et al. 1995) and niche construction theory (Laland et al. 2000; Odling-Smee 2003). Niche construction theory recognizes that the environment changes in response to human natural and cultural selection so that humans "inherit" a change in ecology as well as a change in gene and meme frequency (Laland et al. 2000). We suggest that this "ecological inheritance" is not only the physical and ecological change wrought by people but also the cultural change in response to human activity by animals capable of social learning. Earlier theories were con-

cerned with how these changes fed back to change human genetic and cultural inheritance. We expand this view by suggesting that where human activity results in differential cultural fitness of another animal's memes (cultural selection from humans to the environment), and the resulting cultural evolution in the animal affects the cultural fitness of human memes (cultural selection from the environment to humans), human and animal memes may become coevolved. This cultural coevolution is analogous to traditional genetic coevolution in which reciprocal natural selection among organisms drives mutual change in genes (e.g., crossbill bill depth and conifer cone structure; Benkman 2003). Moose responses to predators were studied by Berger et al. (2001). See note 11, above, for references describing the Orca-seal interaction. Additional details were provided by Glenn Van Blaricom, University of Washington Cooperative Fish and Wildlife Unit.

18. Isack and Reyer (1989). Hussein Isack tells us that the Boran live in the wooded environments of northern Kenya and southern Ethiopia, where they regularly interact with bees and honeyguides. Isack is technically a Gabbra, part of the Oromo ethnic group that includes the Boran. Gabbra live in less wooded habitats in the same region of Africa as the Boran. Gabbra and Boran both follow birds for honey, but Boran culture has specialized honey hunters.

19. Dolphin-human fishing coevolution is reported by Pryor et al. (1990) and Smith (1998).

20. Makahs and crowberries were discussed by Densmore (1939) and reiterated by Gunther (1945).

21. Wells (2002) relates the story of poor folk using sticks from crow nests for firewood.

22. Kurosawa (1994) and Marzluff et al. (1994, 2001) document and summarize corvid increases in cities.

23. Millspaugh et al. (2000) document elk changes in response to hunting.

24. Elliott (1881) reports raven reintroduction by Russians. Goodwin (1978) suggests that Rooks were reintroduced to New Zealand by homesick Europeans. The Hawaiian, Mariana, and White-necked Crows are listed as endangered by the U.S. government. The last Hawaiian Crow alive in the wild was seen July 2002. Tomiałojc (1979) studied crow expansion and effects on Wood Pigeons in Wrocław.

TWO
A Crow Is a Crow, or Is It?

1. Twain (1897).
2. Warne (1926).
3. Smith (1905) discussed ravens and ancient Romans.

4. Quotation from Warne (1926). Body sizes from Clark et al. (1991).

5. Portmann and Stingelin (1961) and Pearson (1972) provide a detailed analysis of the avian brain. Portmann (1946, 1947) measured hundreds of bird brains. Heinrich (1999) reports and discusses some of the amazing intellect displayed by ravens but incorrectly claims that ravens have the largest encephalization quotient (EQ) of all birds measured by Portmann. The ratio of brain hemisphere size to body size (Portmann's EQ) for ravens was 19. EQ for two large macaws measured by Portmann was 28. To measure the relative EQ among corvids, we compared the deviation of each species' brain size from the average regression of brain to body size for eight corvid species. These deviations, known as residuals, suggest that American Crows (not ravens) had the largest relative brain size within corvids (see graph on page 43). Seed caching by nutcrackers and jays is detailed by Vander Wall and Balda (1981) and Vander Wall (1982, 1990). Pinyon Jay nest placement is from Marzluff (1988). String pulling by corvids was reported by Homberg (1957) and investigated in detail by Heinrich (1999).

6. The Avian Brain Nomenclature Consortium (2005) discusses and provides images of avian and mammalian brain structures. Changes in hippocampus volume and use of brain for spatial memory are reported by Harvey and Krebs (1996), Krebs et al. (1996), Clayton (1998), and Pepperberg (1999). Some doubt has since been cast on change in hippocampus volume and the absolute link between hippocampus size and spatial memory abilities (Healy et al. 2005). Emery and Clayton (2004) discuss the corvid forebrain and its use in learning in memory.

7. Angell (1978); Heinrich (1999).

8. Prior et al. (2000) document magpie self-awareness. Paz-y-Miño et al. (2004) show that Pinyon Jays are able to know the rank of new individuals in social groups by observing their interactions with others of known rank.

9. Bendire (1895) relates the story of Jim. Heinrich (1999) reports on insight in ravens.

10. Emery and Clayton (2004) review the mentality of crows and compare them with apes. They introduce the idea of a cognitive tool kit.

11. Implications of black plumage are discussed by Heppner (1970) and Goodwin (1978, 1986).

12. Savage (1995) retells the Peacock story. Sax (2003) reports the Greek legend.

13. Albinism is reported and discussed by Bent (1946), Henderson (1982), Bancroft (1993), and Ogilvie (2003). Predator selection of odd prey is considered by Mueller (1968, 1971) and Curio (1976).

14. Marzluff and Balda (1992) document social recognition by vocalizations.

15. We suspect that all corvids form lifelong pair bonds, but we know this for certain in only a few species. Several jays have been studied for decades, including the in-

dividual marking of thousands of birds to document pair-bonding (Marzluff et al. 1996). Long-term studies are needed to determine which birds pair with which partners and for how long. In addition, genetic studies are needed to confirm that mated partners are genetic parents of nestlings. Long-term observational and genetic studies of crows and ravens are lacking. The insecticidal properties of cedar-lined crow nests was experimentally investigated and shared with us by Professor Peter Arcese (University of British Columbia). Arcese also speculated that native peoples may have learned about cedar by watching crows. Tyler (1979) relates the Pueblo legend.

16. Physical changes with age were measured by Emlen (1936) and Clark et al. (1991).

17. Pittaway (1988) points out wing-tail flicking differences between crows and ravens.

18. House Crows dispersing on boats was reported by Goodwin (1978, 1986).

19. Goodwin (1986) suggested that ancestral Asian crows evolved separately on the mainland and islands.

20. Brooks (1942); Sutton (1951); Davis (1958); Johnston (1961). Sutton was at the University of Michigan at this time, although his ornithological career was spent largely at the University of Oklahoma.

21. Northwestern Crow taxonomy was noted by Suckley and Cooper (1860) and Bendire (1895) but investigated in detail by Johnston (1961).

22. Interbreeding between crows and ravens was documented by Jefferson (1991, 1994).

23. American Crow races are described by Bent (1946) and Verbeek and Caffrey (2002). Florida crow sociality was studied by Kilham (1989).

24. Ohio roosts were studied by Good (1952). Urban roosting was studied by Gorenzal et al. (1996).

25. Banko et al. (2002) reviewed 'Alala biology. The number of 'Alala in captivity in January 2005 was provided by Dr. Scott Derrickson, National Zoological Park, Washington, DC, and Mr. Jay Nelson, U.S. Fish and Wildlife Service, Honolulu.

THREE

Intertwined Ecologies and Mutual Destinies

1. The 8-million-year head start assumes that protocorvids did not develop until the end of Australian isolation 15 million years ago and that humans diverged from other apes 7 million years ago (Erickson et al. 2002; Diamond 1997). It is possible that protocorvids were on the scene 50 million years ago. Human brains have gone through two growth spurts—one about 2 million years ago and the other about 250,000 to 500,000 years ago (Pickett 1988; Arello and Wheeler 1995). Before that

they were unremarkable for primates (see *Australopithecus* in Pickett 1988, fig. 16, or Arello and Wheeler 1995, fig. 6).

2. Gondwanan origins of birds are summarized by Edwards and Bolles (2002) and Ericson et al. (2002).

3. Various ideas on when people colonized North America are from Nemecek (2000) and Roosevelt (2000).

4. Songbird colonization is discussed by articles in note 2, above. Fleischer and McIntosh (2001) investigate Cytochrome *b* differences in corvids. Pielou (1979) describes vegetative aspects of the Bering Land Bridge.

5. Emslie (1998, in press) reports on fossil crows. The lack of crows in the Southwest is from Richards (1971) and Sagebiel (1998).

6. Omland et al. (2000) discovered raven clades. Leonard et al. (2002) use DNA evidence to suggest that humans crossed the Bering Land Bridge twelve thousand to fourteen thousand years ago with dogs domesticated from European wolves.

7. Driver (1999) reports on raven fossils and human settlements from the Pleistocene in Canada.

8. Ratcliffe (1997) details changing European attitudes toward ravens. Cave images can be seen in Armstrong (1958) and Gore (2000). Laws protecting corvids are described by Gurney (1921) and Sax (2003). Fox-Davies (1986) provides illustrations of crow and raven heraldry.

9. The importance of the London fire and the views of Muslims toward corvids is from Sax (2003).

10. The cultural gap in European raven distribution was mapped by Schultz-Soltau (1962).

11. Early American bounties and laws are discussed by Sax (2003).

12. Crow roost destruction was documented by Kalmbach (1939) and Bent (1946).

13. Crow hunting guides include Woodward (1949) and Popowski (1962).

14. Urban land-cover change is measured by Gillham (2002).

15. Crow increases are measured by Marzluff et al. (1994, 2001), McGowan (2001a), Neatherlin (2002), and Withey (2002). Tribe of Crow is printed in Real Change (*Tribe of Crow* 1999).

16. Marzluff et al. (2001) summarizes worldwide increases in corvids. Jungle Crow reports are from Arnold (2000) and Struck (2001).

17. Robinson et al. (2005) document Seattle area land change.

18. Restani et al. (2001) studied raven movements in Greenland.

19. Zwickel and Verbeek (1997) and Verbeek and Caffrey (2002) report maximum lifespans of crows. See also box 2. We predict forty-year-old crows by starting with 100

fledglings, 50 percent juvenile mortality, 25 percent yearling mortality, and 5 percent adult mortality. Fifty will live to one year, 38 to two years (50 × 0.75), 36 to three years (38 × 0.95), 34 to four years (36 × 0.95), 15 to twenty years, and so on, until 5 remain after forty years.

20. Marzluff and Heinrich (1991) showed that, in ravens, territorial defense by adults at rich foods subsided when at least nine vagrants challenged them for access.

21. Withey and Marzluff (2005) detail the study.

22. Malthus (1798). Alberti et al. (2004) document forest conversion in Seattle.

23. Lanciotti et al. (1999) report on West Nile virus. Ridgeway (1893) documented a large die-off of crows in Washington after a severe winter storm. Many had frozen eyes.

24. McGowan (2001a).

FOUR

Inspiration for Legend, Literature, Art, and Language

1. Armstrong (1958).

2. Native American legends described at http://home.no.net/norweagl/lore/lego36. htm; Tyler (1979).

3. Raven creation stories from Genesis 8:7–9, Smith (1905), Armstrong (1958), Coombs (1978), Nelson (1983), Heinrich (1989), Feher-Elston (1991), –Ratcliffe (1997), and Sax (2003). Jewish folklore from Angell (1978).

4. The navigational use of ravens is discussed by Armstrong (1958). Sax (2003) pictures the Celtic helmet adorned with a raven.

5. Boria Sax challenges the traditional legend of ravens in the Tower of London in an unpublished manuscript titled "Ravens in the Tower of London."

6. Raven legends from Boas (1913–1914), Feher-Elston (1991), Savage (1995), and Ratcliffe (1997).

7. Koyukon interactions with ravens are recounted in Nelson (1983).

8. Tyler (1979) recounts the crow coloration story.

9. Legros (1999) quotes McGinty.

10. Coast Salish and Tseshaht crow-raven stories are found in Feher-Elston (1991).

11. Tyler (1979).

12. Savage (1995).

13. Loon's story is found in Armstrong (1998). For other Native American legends, see http://home.no.net/norweagl/lore/lego36.htm, told by Good White Buffalo, South Dakota, 1964.

14. The story of Pet Crow is found at Project Gutenberg (http://www.gutenberg.org):

M. L. McLaughlin, *Myths and legends of the Sioux* (1995), http://digital.library. upenn.edu/webbin/gutbook/lookup?num=341; Tyler (1979).

15. Philip (1997).

16. Crow namesakes are discussed at http://www.angelfire.com/my/rabiddeputydawg/ crowtribe.html and by Anderson (1984).

17. http://www.literature.org/authors/aesop/fables; Lawrence (1997).

18. Sarma (1993).

19. Porter (1909) recounts Pliny's reasoning.

20. Hazelton (1969).

21. Hay (1871).

22. Anderson (2000).

23. Cassidy (1984).

24. Holder (1986); Armstrong (1958).

25. Conversation with Seattle Art Museum staff, 2000.

26. Audubon (1967).

27. Busch's story and illustrations can be found at http://www.rivertext.com/ hans_fr.shtml.

28. For the Association of Old Crows, see http://www.oldcrows.org/Hist.html.

29. King Arthur is discussed by Martin (1993). Other accounts have King Arthur as a chough or raven (Coombs 1978), but all have him as a corvid.

30. Davies (1970).

31. Ravenstone is discussed by Armstrong (1958). Moore (2002) investigated British place names for ravens.

32. Keyes (1998) surveyed human languages for words used to denote crows and ravens.

33. Discussion of Greene (1592) disparaging Shakespeare can be found at http:// ise.uvic.ca/Library/SLT/life/groatsworth2.html.Greene.

34. English proverbs are found in Simpson and Weiner (1989). Wells (2002) tells of "raven's knowledge" and of Jim Crow laws. Sax (2003) expands upon the derivation of Jim Crow.

35. Stein (2000) reprints the photo of an early Makah scarecrow. The Mimbres bowl was first described by Fewkes (1923) as a bird hunter using traditional traps. Brody (1977) published a photograph of the bowl and reinterpreted it as a man trapping birds in a garden. He suggests that it might also be a representation of the Mimbres myth about a hero saving the stars by trapping crows that were eating them. Sax (2003) agrees generally with Brody that the bowl represents crow-trapping in a garden and discussed the evolution of scarecrows. We agree that the bowl repre-

sents crow-trapping, but add that it clearly shows the social learning process of crows. The crosses by the fence were used elsewhere by Mimbres artists to depict corn and are clearly different from the birds' tracks, confirming the setting as a garden, not a typical bird trap.

36. Fergus (1984).

37. Archibald Menzies, in his *Journal of the Vancouver Expedition* of 1792, reported on Northwestern natives eating crow (see Newcombe 1923). Killing and eating ravens are referred to by Nelson (1973) and Salomonsen (1967), but Osgood (1976) reports that ravens were not eaten by the Tanaina. Goodwin (1986) tells of Rook eatings. The cost of crows in Oklahoma is from Kalmbach (1939). Good (1952). Dapkus (2003) details the modern Lithuanian use of crows.

38. Maccarone (1989) studied the sentinel behavior of crows.

FIVE

The Social Customs and Culture of Crows

1. Lorenz (1952) discussed animal communication and emotion. Lorenz (1981) cautions against anthropomorphism.

2. Parr (1997) described the stable and fluid nature of crow society. See Chapter 2, note 15, above, concerning the likelihood that most corvids form lifelong monogamous pair-bonds.

3. Heinrich (1989) details the social life of vagrant ravens.

4. Bailey (1927) and Stouffer and Caccamise (1991) describe crow movements.

5. Valutis and Marzluff (1999); Whitmore and Marzluff (1998).

6. Seasonal variation in crow sociality is reported in Good (1952), Haase (1963), Stouffer and Caccamise (1991), Parr (1997), and Caccamise et al. (1997).

7. Crow sociality has been studied by Caffrey (1991, 1992), Parr (1997), and Withey (2002).

8. Greenwood (1988).

9. McGowan (2001a).

10. Brown (1987) details cooperative breeding.

11. Helping by American and Northwestern Crows is discussed by Verbeek and Butler (1981), Kilham (1984, 1989), Chamberlin-Auger et al. (1990), Caffrey (1991, 1992, 1999, 2000b), Parr (1997), and McGowan (2001a). Possible helping by ravens is reported by Boarman and Heinrich (1999).

12. Theories of helping in crows from Parr (1997) and McGowan (2001a). Of ten nests in one Washington study area in 2002, three were successful, and two of these had helpers. No nests with helpers failed.

13. Baglione et al. (2003).

14. California crow sociality from Caffrey (2000b). Other western crow sociality from Emlen (1942), Johnston (1961), Richards and White (1963), and Butler et al. (1984). Theory of dispersal from Koenig and Pitelka (1979), Greenwood (1980), Woolfenden and Fitzpatrick (1984), and Marzluff and Balda (1989).

15. Goodwin (1986) details allopreening. Other birds, notably parrots and owls, preen often, but in our experience corvid pairs outpreen them all. Kilham (1985) reports on tool use.

16. Richner (1992) documents how food increases crow reproduction.

17. Kilham (1984f).

18. Kilham (1984f); James (1983).

19. Kilham (1984f).

20. Goodwin (1986).

21. Simmons (1957, 1966) interprets anting for insecticide application. Quammen (1985).

22. Animal play is cataloged by Ficken (1977), Byers (1981), Fagen (1981), Beckoff (1984), and Ortega and Bekoff (1987). Crow play is reported from Good (1952) and Kilham (1984e).

23. Kurosawa (1999).

24. Garner (1978) reports on citizens' issues with large bird roosts.

25. Crow predators from Bent (1946), Goodwin (1986), Long et al. (1987), James and Oliphant (1988), and Robinette and Crockett (1999). Long (1990) and Goss (1905, cited in Hill 1999) observed crows killing predators. In the spring of 2002, a Bald Eagle was recovered dead below a crow nest after apparently suffering repeated blows to its head and back from defensive crows. The eagle was raiding a crow nest when caught in the act by the parents and neighbors. Roberts (1903, cited in Hill 1999) tells of a Peregrine Falcon catching a mobbing crow.

26. Pavey and Smyth (1998).

27. Löhrl (1968) recounts jay mobbing. Slagsvold (1984a,b). Ken Dial, University of Montana Flight Lab, told us about body size and flight performance (November 2003). Lorenz (1952) and Barash (1976) document mobbing of black objects by crows.

28. Royko (1993); Jones (1995); Myers (1998); Balter (2002).

29. Large roosts documented by Bent (1946) and Iams (1972). Theory of roosting from Eiserer (1984), Caccamise and Morrison (1986, 1988), Mock et al. (1988), and Weatherhead (1988). Hamilton (1971) developed the "selfish herd" theory of predation risk.

30. Ward and Zahavi (1973) developed the theory of information centers. Marzluff et al. (1996) confirmed the existence of information centers in raven roosts.

31. Wright et al. (2003) studied raven roosts in Wales. Sonerud et al. (2001) studied Hooded Crow roosts in Norway.

32. Marzluff and Heinrich (1991) showed how nine ravens can overpower a territorial pair. Crow foraging was studied by Morrison and Caccamise (1990), Stouffer and Caccamise (1991), and Caccamise et al. (1997).

33. Hansen et al. 2000; Smedshaug 2000.

34. Bent (1946) and Good (1952).

35. Marzluff observed ravens entering and leaving roosts, foraging, and responding to food bonanzas he provided in the sagebrush of southwestern Idaho over a three-year period. Roost departures were unsynchronized, and the accumulation of ravens at carcasses rarely suggested recruitment.

36. Goodwin (1986).

37. Personal communication with Kurt Kotrschal, July 20, 2004.

SIX

Communication and Culture

1. Definition of language from Davies (1970). The vocal behavior of crows and ravens is exceedingly complex and not particularly well understood. What we do know about it, however, suggests some of the properties that strict behaviorists require before an animal's communication system is called a "language." First, there are many unique calls with distinct meaning that appear to be combined at times to communicate increasingly complex information. Fundamentally, calls of various types are used to represent important aspects of a crow's environment symbolically. For example, specific caws are given to attract or to repel other crows. Caws are thus symbolic representations of concepts, just as "Come here" and "Go away" are symbolic representations of concepts to English-speaking people. Some calls may also develop only in specific locales and be acquired by crows by cultural transmission, so it is possible that different crows might have different calls for the same stimulus, much as people have different words for the same concept. Many scientists would not call crow vocalizations a language until it could be demonstrated that symbolic representations were combined using specific cultural rules (syntax) and that symbols were created and showed consistent meaning even if given in novel situations. Richards and Thompson's (1978) paper (see also Thompson 1982) demonstrates the importance of syntax to crow communication. Premack (2004) claims that even rudimentary syntax is not enough to claim the occurrence of true "language." He contends that only humans have language because we have such an infinite ability to create and change meaning by combining words in so-

cially acceptable and grammatically correct ways. We agree that human language is clearly the most unique and complex communication system known, but we disagree that it is fundamentally different from the communication systems of crows and ravens. We view communication systems as more graded in complexity and believe that the evidence suggests that corvid communication systems contain the basic requirements of a language, including basic syntax, so we cautiously use the term *language*. Those interested in this debate would enjoy Hockett (1959), Griffin (1992), and Fitch and Hauser (2004). Crow, raven, and jay details from Marzluff and Heinrich (1991), Marzluff and Balda (1992), and Parr (1997).

2. Heinrich and Marzluff (1991); Marzluff and Heinrich (1991); Enggist-Dueblin and Pfister (2002).

3. Enggist-Dueblin and Pfister (2002).

4. Chamberlain and Cornwell (1971), Reaume (1988), and Parr (1997) describe crow call types.

5. Crow caws and koaws were studied by Frings and Frings (1957) and Chamberlain and Cornwell (1971). The response of French corvids is from Frings et al. (1958).

6. Chaney and Seyfarth (1990) studied vervets. Parr (1997) investigated crow calls.

7. Individual recognition using vocalizations by a corvid, the Pinyon Jay, was studied by Marzluff and Balda (1992).

8. Chamberlain and Cornwell (1971) described the crow scream, which is commercially available from Johnny Stewart Crow Calls (1971 and 1992, Stewart Outdoors). The cultural evolution of predator recognition is found in Curio (1976, 1978) and Curio et al. (1978).

9. Vocal recognition of mates in Pinyon Jays by Marzluff (1988a).

10. Bent (1946) quotes Townsend.

11. Brown (1985) and Brown and Farabaugh (1997) studied crow song dialects.

12. Mimicry is reported in Chamberlain and Cornwell (1971) and Byers (1990).

13. Lorenz (1952) relates the story of Hansl and Roah. Savage (1995) reports a crow mimicking a lost mate.

14. Brown (1983); Parr (1997).

15. Thompson (1968, 1982); Chamberlain and Cornwell (1971); Brown (1983); Parr (1997).

16. Brown (1985) comments on changing calls with predator movements.

17. Parr (1997) documents doubling cadence of territory defense.

18. Bossema and Benus (1985) documents pincer tactics.

19. Wilson (1971) describes insect communication.

20. The abilities of dogs to read our visual signaling, including only our gaze, is demonstrated by Hare et al. (2002).

21. Thompson (1968, 1982).

22. Pepperberg (1999) discusses counting in birds. Koehler (1950) describes his experiments.

23. Bugnyar and Kotrschal (2002) demonstrate caching deception by ravens.

24. Steller's Jays often mimic hawks (Hope 1980), and we suspect that they do this to distract other jays or other bird species they forage with, especially in winter, when they often flock together. Munn (1986) details deception in shrike-tanagers and antshrikes.

SEVEN

Reaping What We Sow

1. Pimm (2001) documents forest cover loss in the eastern United States.

2. Houston (1977, 1980) reports on changing patterns of corvids on the Canadian prairies. Brewer et al. (1991) and Sharpe et al. (2001) document changes in Michigan. Marzluff et al. (1994) measured crow changes in the western United States.

3. Kalmbach (1939) documents crow migratory behavior. Houston (1969) reports on crow movements from Canada and notes that documenting the migratory habits of crows has been challenging. Biologists banded (ringed) thousands of crows in the north, hoping that others would catch or kill them and report their findings. Although this did happen, the results were not without incident. In Canada, Fred Bard banded crows to learn more. Canadian wildlife managers banded crows to reduce their numbers. Fearful that crows were eating too many ducklings, the mangers applied "reward bands" to encourage hunting. Hunters targeted banded crows in the hopes of getting a valuable cash reward or nifty prize. Incidentally, they shot a lot of Bard's crows. To save his scientific investigations, Bard painted his crows' bands black. Fewer hunters shot them, and we all learned more about crow movements and behavior.

4. Kalmbach (1939) dissected crow stomachs. Quiring and Timmins (1988) and Solem (1997) documented crows eating pest insects.

5. Crow diets from Bent (1946), Good (1952), Kilham (1982a,b,c, 1984a,b,d, 1985), Cuccia (1984), DiLabio and Dunn (1985), Maccarone (1991), Septon (1991), Putnam (1992), Marzluff et al. (2001), and Marzluff and Neatherlin (in press). Kilham (1985b) described pig riding.

6. Reaume (1987).

7. Marzluff et al. (2001); Marzluff and Neatherlin (in press).

8. Heinrich (1988, 1999) and Heinrich et al. (1995) describe jumping jacks and ravens' fear of food. Ward and Low (1997) measured crow vigilance in urban areas. Knight et al. (1991) and Skagen et al. (1991) studied corvid-eagle interactions with salmon.

9. Mayr (1974) discussed general attributes of social animals. Robinette and Ha (2000) and Szpir (2003) report on scrounging by Northwestern Crows.

10. Abilities of seed-caching corvids are detailed in Balda (1980), Vander Wall (1990), Marzluff and Balda (1992), and Balda, Pepperberg, and Kamil (1998).

11. Kilham (1984d) discussed American Crow caching. James and Verbeek (1983, 1984, 1985) document Northwestern Crow caching.

12. Reineke (1995) details the crow caching experiments. Balda (1980) and Vander Wall (1982) report accuracies of jays and nutcrackers recovering caches.

13. Heinrich (1999) documents doughnut caching by ravens.

14. Heinrich and Pepper (1998) studied raven caching. Emery and Clayton (2004) discuss evidence from jays that suggest the ability to ignore caches of perishable, but not lasting, items.

15. Seaton (1898).

16. Smith (1989) describes golf ball stealing. WhiteBoard News (1997) reports on the loss of the gold bracelet.

17. Corvid nest predation has been studied by Sugden (1987), Sullivan and Dinsmore (1990), Ewins (1991), Marzluff and Balda (1992), Freeman (1993), Clark et al. (1995), and Vigallon and Marzluff (2005). Gotmark et al. (1990) document crows watching loons. Sonerud and Fjeld (1987) document crows rechecking previously used nest boxes.

18. Egg choice was documented by Montevecchi (1976). Schauer and Murphy (1996) and Rossow (1999) report on egg predation by ravens on murres.

19. Montevecchi (1976); Heinrich et al. (1995); Marzluff et al. (2000); Luginbuhl et al. (2001).

20. Nest predation cases are reported from Kalmbach (1939), Ewins (1991), Freeman (1993), and Dickinson (2003). Marzluff's work is detailed in Marzluff et al. (2000), Luginbuhl et al. (2001), Neatherlin and Marzluff (2004), and Bradley and Marzluff (2003). Duck nesting after crow removal was followed by Clark et al. (1995).

21. Video studies by Thompson et al. (1999) and Pietz and Granfors (2000).

22. Kristan and Boarman (2003).

23. Paine (1966) presents the starfish example. Stoffel (2002) records owls using old raven nests, and Houston (1977) discusses Merlin use of crow nests.

24. Worldwide increases in corvids from Eden (1985), Fraissinett (1989), Marzluff et

al. (1994), Konstantinov (1996), Hogrefe and Yahner (1998), Haskel et al. (2001), Jerzak (2001), and Marzluff et al. (2001). Effects on endangered species from Boarman (1993) and Miller et al. (1997).

25. Kilham (1982a).

26. Corvid food dropping detailed by Takagi and Ueda (2002). Montevecchi (1978) documents Fish Crows and ravens dropping objects on gulls.

27. Cristol and Switzer (1999) document walnut dropping.

28. Nihei (1995) and Nihei and Higuchi (2001) describe and photograph nut-cracking. Hito Higuchi, Lab of Biodiversity Science, University of Tokyo, told us about driver attitudes (personal communication, November 8, 2003).

29. Zach (1978, 1979).

30. Homberg (1957) reports on ice-fishing crows. Powell and Kelly (1977), Savage (1995), and Caffrey (2000, 2001) document tool use by American Crow. Hunt (1996, 2000), Hunt et al. (2001), and Weir et al. (2002) studied tool use by New Caledonian Crows. Hunt (2000) relates crow tool use to early human tool use.

31. Raven and crow troubles from *Offbeat News* (2003) and *Mainichi Daily News* (2003).

32. Arnold (2000); Associated Press (2001); Soh et al. (2002); Brook et al. (2003).

33. Dickinson (2003).

34. Proceedings of the Wild Bird Society of Japan symposium, "How Should We Deal with the Crow Problems in Tokyo," held October 9, 1999.

35. Struck (2001); AFP-jiji (2002).

36. Shirota (1989); AFP-jiji (2002); Connell (2003).

37. The entire sequence of news articles on the Chatham Crow Wars is archived at http://www.netrover.com/~rsiddall/DailyNews.html.

38. http://www.riverdeep.net/current/s001/02/020201_crows.jhtml.

EIGHT

Centering the Balance

1. Fossil crows were uncovered by James and Olson (1991) and Olson and James (1991). Rob Fleisher's recent genetic analysis of these crows confirms Olson and James's suspicion that the smaller specimen from Maui is 'Alala. Counts of 'Alala as of January 2005 were provided by Dr. Scott Derrickson, National Zoological Park, Washington, DC, and Jay Nelson, U.S. Fish and Wildlife Service, Honolulu.

2. Darwin (1859); Wallace (1881); Darwin and Wallace (1858). Modern extinctions tabulated by Pimm et al. (1994).

3. Hardy (1960) tells of sailors dumping water with mosquitoes into a stream on Maui. Scott et al. (1986) also discuss the arrival of mosquitoes and the possible

routes malaria traveled to reach Hawaii. Van Riper and Scott (2001) document malaria and pox effects.

4. Duckworth et al. (1992) discuss conservation issues, and Banko et al. (2002) detail basic biological and political aspects of the 'Alala.

5. The 2003 draft recovery plan for the 'Alala by the U.S. Fish and Wildlife Service estimates that more than eleven million dollars will be needed over five years to restore the 'Alala to the wild. The plan predicts that costs will remain at that level or increase slightly for each of the next several decades.

6. Wiles et al. (2003) report effects of snakes on birds. Duckworth et al. (1997) document conservation needs of the 'Alala. Tino Augon, Guam Department of Agriculture and Wildlife Resources (personal communication), reports that as of October 8, 2003, there were ten Aga in the wild (six from releases in 2000 and 2001 and four from releases in 2002 and 2003).

7. Tarr and Fleischer (1999). Aga on Rota and Guam are genetically too similar to estimate accurately how long they have been separated. Dinerstein (2003) discusses bottlenecks and the importance of keeping species in the ecological mix.

8. Johnston (1961) measured Northwestern Crows and American Crows in the Pacific Northwest. Dawson and Bowles (1909) observed unique aspects of Northwestern and American Crows in the Puget lowlands. Fran James (1983), Florida State University, points out that coastal populations of a species are typically smaller than those found in the interior, and often colder, reaches of a continent.

9. Basics of West Nile virus in Lanciotti et al. (1999), Komar et al. (2001), Hall (2003), and http://www.cdc.gov/ncidod/dvbid/westnile.

10. Komar et al. (2003) tested effects of West Nile virus on crows. This article and M. Hutchinson, Pennsylvania Department of Environmental Protection (personal communication), document how West Nile virus is passed among crows. Caffrey (2003), Caffrey and Peterson (2003), Yaremych (2003), Yaremych et al. (2004), Caffrey et al. (2005), and Kevin McGowan (personal communication) measured effects of West Nile virus on marked, wild American Crow populations. The most recent study by Caffrey et al. (2005) documents crow losses over two successive years of West Nile virus exposure and concludes that 72 percent of an Oklahoma population of crows was lost, including 82 percent of all juveniles born during 2002 and 2003.

11. Mashimo et al. (2002) document genetic variation in mouse susceptibility to West Nile virus. Dawkins (1986) draws the analogy between natural selection and a blind watchmaker. Genetic and cultural change are intimately linked, as suggested by the theory of gene-culture coevolution (Durham 1991) and niche construction (Odling-Smee et al. 2003). Cultural responses of crows to West Nile virus that en-

hance an individual's fitness may become more strongly genetically determined, restricting cultural expression to the most adaptive behaviors within tens of generations (Lachlan and Feldman 2003).

12. Glandt (1991); Glandt and Conrad (2001).

13. Sauer (2003) documents how people resist lethal control of crows.

14. Neatherlin and Marzluff (2004) show how crows use campgrounds. Tomback et al. (1990) suggest that park handouts may decrease seed caching by Clark's Nutcrackers.

15. Hanson (1946) reports on the effects of killing roosting crows.

16. Tino Augon, Guam Department of Aquatic and Wildlife Resources, told us how many Aga were released (personal communication, December 2003).

17. Whitmore and Marzluff (1998); Valutis and Marzluff (1999).

18. These costs apply to vertebrate species only and are highly influenced by a few very well funded species. Others get even less support. Restani and Marzluff (2002).

19. Wilson (1984) proposed the biophilia hypothesis. Cook (1771) noted Hawaiians keeping crows in their villages. Thoreau's March 4, 1859, entry from his *Journal* is recounted by Cruickshank (1964).

NINE
Future Interactions

1. Eskimo use of ravens was reported by Freuchen and Salomonsen (1958). Heinrich (1999) investigated this belief but in his short visit to the Arctic was unable to confirm it. Isack and Reyer (1989) describe the honeyguide-Boran relationship.

2. Dawson and Bowles (1909) report on taboos.

3. Blumenschine (1987) and Blumenschine et al. (1994) discuss passive scavenging by early humans, Bunn (1996) discusses early humans overpowering carnivores, and Tunnell (1996) and Rose and Marshall (1996) discuss human defense of food from carnivores. Occurrence of African corvids from Madge and Burn (1994). Rancho La Brea fossils are described by Howard (1962).

4. Stiner (2002) details the early hunting activities of humans. Richards et al. (2000) confirm carnivory in European Neanderthals.

5. Associations of ravens with wolves are discussed by Heinrich (1999). The potential effects of raven scavenging on wolf sociality are proposed by Vucetich et al. (2004).

6. In 2002 one of our radio-tagged crows was killed by Quinault tribal members as it raided their fish holding pens. This was one of only three tagged crows to die during our study. Nelson (1983) described current interactions between ravens and Koyukon.

7. Sax (2003) discusses crows in coats of arms and Fox-Davies (1986) provides illustrations.

8. Ratcliffe (1997) describes European persecution of corvids. Recent reintroductions of ravens are reported by Renssen (1988), Glandt (1991), and Conrad and Glandt (2001).

9. United Nations (1999) projects urban growth. Raven use of human artifacts in deserts and shrublands from Steenhof et al. (1993) and Knight et al. (1993). Urban crow roosts are explored by Gorenzel and Salmon (1995) and Gorenzel et al. (1996). Fossil crows in ancient human settlements are reported by Wyrost (1993), Krönneck (1995), Dobney et al. (1999), Driver (1999), and Cavallo (2000).

10. Marzluff and Heinrich (1991) measured raven defense of food. Ravens at dumps are described by Conner and Adkisson (1976), White and White (1988), Huber (1991), Knight et al. (1995), Restani et al. (1996, 2001) and Skarphédinsson et al. (1998). McGowan (2001a) studied dispersal and family size in crows.

11. Nihei and Higuchi (2001).

12. Cultural carrying capacities are measured by Calhoun (1952), Organ and Ellingwood (2000), and West and Parkhurst (2002). Reuters (2003) reports on the drunk crow. Young and Engels (1985) describe the fecal shields on Idaho power lines. Hito Higuchi, Lab of Biodiversity Science, University of Tokyo, told us about human responses to crows cracking nuts (personal communication, November 8, 2003).

13. Goodwin (1978) reports Rooks being brought from England to New Zealand.

14. Crow hunting is described by Woodward (1949), Popowski (1962), http://www.crowbusters.com, and Reid Hargiss (personal communication, August 12, 2003). Dapkus (2003) tells of Lithuanians' newfound fondness of crow.

15. Trends in bird feeding from Shaw and Mannan (1984), Boyle and Samson (1985), and http://www.birdfeeding.org/media.html. Alverdes's story was reported in the November 24, 2003, *Seattle Times*.

16. The concept of culture as a complex whole, including knowledge, belief, art, morals, customs, habits, and capabilities, was first articulated by Tylor (1871).

17. Animal domestication is discussed by Diamond (1998) and Price (2002).

18. Gilbert (1988); Byers (1990).

19. On the social benefits of nature see Kaplan and Kaplan (1989), McElroy (1996), Dannenberg et al. (2003), and http://www.deltasociety.org/dcs000.htm.

20. Thoreau (1884).

21. Nelson (1983).

References

AFP-jiji. 2002. Ishihara losing war against feathered foes. *Japan Times* October 17.

Aiello, L. C., and P. Wheeler. 1995. The expensive-tissue hypothesis. *Current Anthropology* 36:199–221.

Alberti, M., R. Weeks, and S. Coe. 2004. Urban land cover change analysis for the central Puget Sound: 1991–1999. *Journal of Photogrammetry and Remote Sensing* 70:1043–1052.

Altshuler, D. L., and C. J. Clark. 2003. Darwin's hummingbirds. *Science* 300:588–589.

Anderson, D. 2000. *Blues for unemployed secret police: Poems*. Willimantic, CT: Curbstone.

Angell, T. 1978. *Ravens, crows, magpies, and jays*. Seattle: Univ. of Washington Press.

Armstrong, E. A. 1958. *The folklore of birds*. London: Collins.

Arnold, W. 2000. A fastidious city-state has an answer for crows. *New York Times International* September 21.

Audubon, J. J. 1967. *The birds of America*, vol. 4. New York: Dover.

Avian Brain Nomenclature Consortium. 2005. Avian brains and a new understanding of vertebrate brain evolution. *Nature Reviews: Neuroscience* 6:151–159.

Avital, E., and E. Jablonka. 2000. *Animal traditions: Behavioural inheritance in evolution*. Cambridge: Cambridge Univ. Press.

Baglione, V., D. Canestrari, J. M. Marcos, and J. Ekman. 2003. Kin selection in cooperative alliances of Carrion Crows. *Science* 300:1947–1948.

Bailey, A. M. 1927. Notes on the birds of southeastern Alaska. *Auk* 44:351–455.

Balda, R. P. 1980. Recovery of cached seeds by a captive *Nucifraga caryocatastes*. *Zeitschrift für Tierpsychologie* 52:331–346.

Balda, R. P., I. M. Pepperberg, and A. C. Kamil, eds. 1998. *Animal cognition in nature*. London: Academic.

Balter, J. 2002. Nothing to crow about. *Seattle Times* June 27.

Bancroft, J. 1993. Further observations of albinism in birds. *Blue Jay* 51:203–205.

Banko, P. C., D. L. Ball, and W. E. Banko. 2002. Hawaiian crow. *Birds of North America* 648:1–28.

Barash, D. P. 1976. Mobbing behavior by crows: The effect of the "crow-in-distress" model. *Condor* 78:120.

Barker, F. K., G. F. Barrowclough, and J. G. Groth. 2002. A phylogenetic hypothesis for passerine birds: Taxonomic and biogeographic implications of an analysis of nuclear DNA sequence data. *Proceedings of the Royal Society (London)* 269:295–308.

Barrett-Lennard, L. G., J. K. B. Ford, and K. A. Heise. 1996. The mixed blessings of echolocation: Differences in sonar use by fish-eating and mammal-eating killer whales. *Animal Behaviour* 51:553–565.

Bekoff, M. 1984. Social play behavior. *BioScience* 34:228–233.

Bendire, C. 1895. *Life histories of North American birds, from the parrots to the grackles, with special reference to their breeding habits and eggs*. Smithsonian Institute Special Bulletin. Washington, DC: Government Printing Office.

Benkman, C. W. 2003. Divergent selection drives the adaptive radiation of crossbills. *Evolution* 57:1176–1181.

Bent, A. C. 1946. *Life histories of North American jays, crows, and titmice, part 2*. New York: Dover.

Berger, J., J. E. Swenson, and I. L. Persson. 2001. Recolonizing carnivores and naïve prey: Conservation lesions from Pleistocene extinctions. *Science* 291:1036–1039.

Blackmore, S. 1999. *The meme machine*. Oxford: Oxford Univ. Press.

Blumenschine, R. J. 1987. Characteristics of an early hominid scavenging niche. *Current Anthropology* 28:383–407.

Blumenschine, R. J., J. A. Cavallo, and S. D. Capaldo. 1994. Competition for carcasses and early hominid behavioral ecology: A case study and conceptual framework. *Journal of Human Evolution* 27:197–213.

Boarman, W. I. 1993. When a native predator becomes a pest: A case study. In *Conservation and resource management,* ed. S. K. Majumdar, E. W. Miller, D. E. Miller, E. K. Brown, J. R. Pratt, and R. F. Schmalz, 191–206. Philadelphia: Pennsylvania Academy of Science.

Boas, F. 1913–1914. Thirty-fifth Annual Report of the Bureau of American Ethnology, Part 1:606.

Bonner, J. T. 1980. *The evolution of culture in animals.* Princeton, NJ: Princeton Univ. Press.

Bossema, I., and R. F. Benus. 1985. Territorial defence and intra-pair cooperation in the Carrion Crow (*Corvus corone*). *Behavioral Ecology and Sociobiology* 16:99–104.

Boyd, R., and P. J. Richerson. 1985. *Culture and the evolutionary process.* Chicago: Univ. of Chicago Press.

Boyle, S. A., and F. B. Samson. 1985. Effects of nonconsumptive recreation on wildlife: A review. *Wildlife Society Bulletin* 13:110–116.

Bradley, J. E., and J. M. Marzluff. 2003. Rodents as nest predators: Influences on predatory behavior and consequences to nesting birds. *Auk* 120:1180–1187.

Bratingham, P. J. 1998. Hominid-carnivore coevolution and invasion of the predator guild. *Journal of Anthropological Archaeology* 17:327–353.

Brewer, R., G. A. McPeek, and R. J. Adams. 1991. *The atlas of breeding birds of Michigan.* East Lansing: Michigan State Univ. Press.

Brody, J. J. 1977. *Mimbres painted pottery.* Santa Fe, NM: School of American Research.

Brook, B. W., N. S. Sodhi, M. C. K. Soh, and H. C. Lim. 2003. Abundance and projected control of invasive House Crows in Singapore. *Journal of Wildlife Management* 67:808–817.

Brooks, A. 1942. The status of the Northwestern Crow. *Condor* 44:166–167.

Brown, E. D. 1983. Functional and adaptive significance of mobbing and alarm calls of the Common Crow *(Corvus brachyrhynchos).* Ph.D. diss., Univ. of Maryland, College Park.

———. 1985. The role of song and vocal imitation among Common Crows (*Corvus brachyrhynchos*). *Zeitschrift für Tierpsychologie* 68:115–136.

Brown, E. D., and S. M. Farabaugh. 1997. What birds with complex social relationships can tell us about vocal learning: Vocal sharing in avian groups. In *Social influences on vocal development,* ed. C. T. Snowdon and M. Hausberger, 98–127. Cambridge: Cambridge Univ. Press.

Brown, J. L. 1987. *Helping and communal breeding in birds.* Princeton, NJ: Princeton Univ. Press.

Bugnyar, T., and K. Kotrschal. 2002. Observational learning and raiding of food caches in ravens, *Corvus corax:* Is it "tactical" deception? *Animal Behaviour* 64:195.

Bugnyar, T., M. Stowe, and B. Heinrich. 2004. Ravens, *Corvus corax*, follow gaze direction of humans around obstacles. *Proceedings of the Royal Society of London (B)* 271:1331–1336.

Bunn, H. T. 1996. Reply to Rose and Marshall. *Current Anthropology* 37:321–323.

Busch, W. 1867. Hans Huckebein, der Unglücksrabe. Originally published as a series in *Die illustrierte Welt.* Available online at http://www.rivertext.com/hans_fr.shtml.

Butler, R. W., N. A. M. Verbeek, and H. Richardson. 1984. The breeding biology of the Northwestern Crow. *Wilson Bulletin* 96:408–418.

Byers, J. A. 1981. The significance of play. *Science* 212:1493–1494.

Byers, P. 1990. *The lost folk art of crow taming.* Quaker City, OH: Seneca Lake Bird and Field Naturalists Club.

Caccamise, D. F., and D. W. Morrison. 1986. Avian communal roosting: Implications of diurnal activity centers. *American Naturalist* 128:191–198.

———. 1988. Avian communal roosting: A test of the "patch-sitting" hypothesis. *Condor* 90:453–458.

Caccamise, D. F., L. M. Reed, J. Romanowski, and P. C. Stouffer. 1997. Roosting behavior and group territoriality in American Crows. *Auk* 114:628–637.

Caffrey, C. 1991. Breeding group structure and the effect of helpers in cooperatively breeding western American Crows. Ph.D. diss., Univ. of California, Los Angeles.

———. 1992. Female-biased delayed dispersal and helping in American Crows. *Auk* 109:609–619.

———. 1999. Feeding rates and individual contributions to feeding at nests in cooperatively breeding western American Crows. *Auk* 116:836–841.

———. 2000a. Tool modification and use by an American Crow. *Wilson Bulletin* 112:283–284.

———. 2000b. Correlates of reproductive success in cooperatively breeding western American Crows: If helpers help, it's not by much. *Condor* 102:333–341.

———. 2001. Goal-directed use of objects by American Crows. *Wilson Bulletin* 113:114–115.

Caffrey, C., and C. C. Peterson. 2003. Christmas bird count data suggest West Nile virus may not be a conservation issue in northeastern United States. *American Birds* 103:14–21.

Caffrey, C., S. C. R. Smith, and T. J. Weston. 2005. West Nile virus devastates an American Crow population. *Condor* 107:128–132.

Caffrey, C., T. J. Weston, and S. C. R. Smith. 2003. High mortality among marked crows subsequent to the arrival of West Nile virus. *Wildlife Society Bulletin* 31:870–872.

Calhoun, J. B. 1952. The social aspects of population dynamics. *Journal of Mammalogy* 33:139–159.

Canterbury, R. A., and D. M. Stover. 1992. Observations of communal roosting of American Crows at Beckley, Raleigh County, West Virginia. *Redstart* 59:82.

Carroll, L. 1872. *Through the looking-glass*. New York: William Morrow.

Cassidy, W. L. 1984. Crow augury. http://www.jcrows.com/crolang.html.

Castro, L., and M. A. Toro. 2004. The evolution of culture: From primate social learning to human culture. *Proceedings of the National Academy of Sciences* 101:10235–10240.

Cavallo, C. 2000. *Animals in the steppe*. BAR International Series 891. Oxford: John and Erica Hedges.

Chamberlain, D. R., and G. W. Cornwell. 1971. Selected vocalizations of the Common Crow. *Auk* 88:613–634.

Chamberlain, D. R., W. B. Gross, G. W. Cornwell, and H. S. Mosby. 1968. Syringeal anatomy in the Common Crow. *Auk* 85:244–252.

Chamberlain-Auger, J. A., P. J. Auger, and E. G. Strauss. 1990. Breeding biology of American Crows. *Wilson Bulletin* 102:615–622.

Cheney, D. L., and R. M. Seyfarth. 1990. *How monkeys see the world: Inside the mind of another species*. Chicago: Univ. of Chicago Press.

Cibois, A., and E. Pasquet. 1999. Molecular analysis of the phylogeny of eleven genera of the Corvidae. *Ibis* 141:297–306.

Clark, R. G., P. C. James, and J. B. Morari. 1991. Sexing adult and yearling American Crows by external measurements and discriminant analysis. *Journal of Field Ornithology* 62:132–138.

Clark, R. G., D. E. Meger, and J. B. Ignatiuk. 1995. Removing American Crows and duck nesting success. *Canadian Journal of Zoology* 73:518–522.

Clark, R. M. 1998–2002. Corvus—the crow. http://www.r-clarke.org.uk/constellations/corvus.htm.

Clayton, N. S. 1998. Memory and hippocampus in food-storing birds: A comparative approach. *Neuropharmacology* 37:441–452.

Connell, R. 2003. City eats crow as professor caws over his spicy sack. *Mainichi Daily News* July 7.

Conner, R. N., and C. S. Adkisson. 1976. Concentration of foraging Common Ravens along the Trans-Canada Highway. *Canadian Field-Naturalist* 90:496–497.

Cook, J. 1771. *A journal of a voyage round the world, in His Majesty's Ship "Endeavour,"* *in the years 1768, 1769, 1770, and 1771*. London: T. Becket and P. A. DeHondt.

Coombs, C. J. F. 1978. *The crows*. London: Batsford.

Crins, W. J. 1985. Partial albinism and the determination of local movements in an American Crow (*Corvus brachyrhynchos*). *Ontario Birds* 3:106–108.

Cristol, D. A., and P. V. Switzer. 1999. Avian prey-dropping behavior. II. American Crows and walnuts. *Behavioral Ecology* 10:220–226.

Cruickshank, A. D. 1939. The behavior of some Corvidae. *Bird Lore* 41:78–81.

Cruickshank, H. 1964. *Thoreau on birds*. New York: McGraw-Hill.

Cuccia, J. 1984. American Crow attacks European Starling in mid-air. *Kingbird* 34:32.

Curio, E. 1976. *The ethology of predation*. Berlin: Springer.

———. 1978. The adaptive significance of avian mobbing. I. Teleonomic hypotheses and predictions. *Zietshrift für Tierpsychologie* 48:175–183.

Curio, E., U. Ernst, and W. Wieth. 1978. The adaptive significance of avian mobbing. II. Cultural transmission of enemy recognition in blackbirds: Effectiveness and some constraints. *Zietshrift für Tierpsychologie* 48:199–217.

Danchin, É., L.-A. Giraldeau, T. J. Valone, and R. H. Wagner. 2004. Public information: From nosy neighbors to cultural evolution. *Science* 305:487–491.

Dannenberg, A. L., R. J. Jackson, H. Frumkin, R. A. Schieber, M. Pratt, and H. H. Tilson. 2003. The impact of community design and land-use choices on public health: A scientific agenda. *American Journal of Public Health* 93:1500–1508.

Dapkus, L. 2003. More Lithuanians hunting, eating crow. Associated Press, June 9. http://espn.go.com/outdoors/hunting/news/2003/0612/1566988.html.

Darwin, C. 1859. *On the origin of species*. London: J. Murray.

Darwin, C., and A. R. Wallace. 1858. On the tendency of species to form varieties. *Journal of the Proceedings of the Linnean Society: Zoology* 3:45–62.

Davies, P., ed. 1970. *The American Heritage dictionary of the English language*. New York: Dell.

Davis, L. I. 1958. Acoustic evidence of relationship in North American crows. *Wilson Bulletin* 70:151–167.

Dawkins, R. 1976. *The selfish gene*. Oxford: Oxford Univ. Press.

———. 1986. *The blind watchmaker*. New York: W.W. Norton.

Dawson, W. L., and J. H. Bowles. 1909. *The birds of Washington*. Seattle, WA: Occidental.

Deeke, V. B., J. K. B. Ford, and P. Spong. 2000. Dialect change in resident Killer Whales: Implications for vocal learning and cultural transmission. *Animal Behavior* 60:629–638.

Deeke, V. B., P. J. B. Slater, and J. K. B. Ford. 2002. Selective habituation shapes acoustic predator recognition in Harbor Seals. *Nature* 420:171–173.

Densmore, F. 1939. *Nootka and Quileute music.* Smithsonian Institution Bureau of American Ethnology. Bulletin 124. Washington, DC: Government Printing Office.

Dewar, H. 2003. Did West Nile virus kill Alexander? *Seattle Times* December 13.

Diamond, J. 1998. *Guns, germs, and steel.* New York: W. W. Norton.

Dickinson, D. 2003. An "airborne black tide mark" is menacing the native bird life of East Africa, according to ornithologists. *BBC News UK Edition* November 11.

Dinerstein, E. 2003. *The return of the unicorns: The natural history and conservation of the Greater One-Horned Rhinoceros.* New York: Columbia Univ. Press.

Di Labio, B. M., and P. M. Dunn. 1985. Apparent predation on a bat by an American Crow. *Trail and Landscape* 19:86.

Dobney, K., M. Beech, and D. Jaques. 1999. Hunting the broad spectrum revolution: The characterization of early Neolithic animal exploitation at Qermez Dere, Northern Mesopotamia. In *Zooarchaeology of the Pleistocene/Holocene boundary,* ed. J. C. Driver, 47–57. BAR International Series 800. Oxford: John and Erica Hedges.

Driver, J. C. 1999. Stratified faunas from Charlie Lake cave and the peopling of the western interior of Canada. In *Zooarchaeology of the Pleistocene/Holocene boundary,* ed. J. C. Driver, 77–83. BAR International Series 800. Oxford: John and Erica Hedges.

Duckworth, W. D., S. R. Beissinger, S. R. Derrickson, T. H. Fritts, S. M. Haig, F. C. James, J. M. Marzluff, and B. A. Rideout. 1997. *The scientific bases for preservation of the Mariana Crow.* Washington, DC: National Academy Press.

Duckworth, W. D., T. J. Cade, H. L. Carson, S. R. Derrickson, J. W. Fitzpatrick, and F. C. James. 1992. *Scientific bases for preservation of the Hawaiian Crow.* Washington, DC: National Academy Press.

Durham, W. H. 1991. *Coevolution: Genes, culture, and human diversity.* Stanford, CA: Stanford University Press.

Eden, S. F. 1985. The comparative breeding biology of magpies, *Pica pica,* in an urban and rural habitat. *Journal of Zoology (London) A* 205:325–334.

Edwards, S. V., and W. E. Boles. 2002. Out of Gondwana: The origin of passerine birds. *Trends in Ecology and Evolution* 17:347–349.

Eiserer, L. A. 1984. Communal roosting in birds. *Bird Behaviour* 5:61–80.

Elliott, H. W. 1881. The Seal-islands of Alaska. In *The history and present condition of the fishery industries,* ed. S. F. Baird and G. B. Goode, 126. Washington, DC: Government Printing Office.

Emery, N. J., and N. S. Clayton. 2004. The mentality of crows: Convergent evolution of intelligence in corvids and apes. *Science* 306:1903–1907.

Emlen, J. T. 1936. Age determination in the American Crow. *Condor* 38:99–102.

———. 1938. Midwinter distribution of the American Crow in New York State. *Ecology* 19:264–275.

———. 1940. The midwinter distribution of the crow in California. *Condor* 42:287–294.

———. 1942. Notes on a nesting colony of western crows. *Birdbanding* 13:143–154.

Emslie, S. D. 1998. Avian community, climate, and sea-level changes in the Plio-Pleistocene of the Florida peninsula. *Ornithological Monographs* 50: 73–90.

———. In press. The early and middle Pleistocene avifauna from Porcupine Cave, Colorado. In *Early and Middle Pleistocene biodiversity and environmental change: The Porcupine Cave fauna from Colorado,* ed. A. Barnowsky. Berkeley: Univ. of California Press.

Enggist-Dueblin, P., and U. Pfister. 2002. Cultural transmission of vocalizations in ravens, *Corvus corax. Animal Behaviour* 64:831–841.

Enserink, M. 2002. West Nile's surprisingly swift continental sweep. *Science* 297:1988–1989.

Ericson, P. G. P., L. Christidis, A. Cooper, M. Irestedt, J. Jackson, U. S. Johansson, and J. A. Norman. 2002. A Gondwanan origin of passerine birds supported by DNA sequences of the endemic New Zealand wrens. *Proceedings of the Royal Society (London)* 269:235–241.

Evans, K., J. Rensel, and V. A. Smith, 1997. American Crows nest in Ogden. *Utah Birds* 13:61–63.

Ewins, P. J. 1991. Egg predation by corvids in gull colonies on Lake Huron. *Colonial Waterbirds* 14:186–189.

Fagen, R. 1981. *Animal play behavior.* New York: Oxford Univ. Press.

Feher-Elston, C. 1991. *Ravensong: A natural and fabulous history of ravens and crows.* Flagstaff, AZ: Northland.

Feldman, M. W., K. Aoki, and J. Kumm. 1996. Individual versus social learning: Evolutionary analysis in a fluctuating environment. *Anthropological Science* 104:209–231.

Fergus, C. 1984. *The wingless crow.* New York: Lyons and Burford.

Fewkes, J. W. 1923. Designs on prehistoric pottery from the Mimbres Valley, New Mexico. *Smithsonian Miscellaneous Collections* 74(6):1–47.

Ficken, M. S. 1977. Avian play. *Auk* 94:573–582.

Fitch, W. T., and M. D. Hauser. 2004. Computational constraints on syntactic processing in a nonhuman primate. *Science* 303:377–380.

Fleischer, R. C., and C. E. McIntosh. 2001. Molecular systematics and biogeography of the Hawaiian avifauna. *Studies in Avian Biology* 22:51–60.

Fox-Davies, A. C. 1986. *The art of heraldry.* London: Bloomsbury Books.

Fraissinet, M. 1989. Espansione della taccola, *Corvus monedula,* nei capoluoghi italiani. *Rivista Italiana di Ornitologia* 59:33–42.

Frazier, J. 2001. Count on crows. *Utne Reader* 104:33–35.

Freeman, J. 1993. American Crow predation on nestlings in Carter County, Oklahoma. *Bulletin of the Oklahoma Ornithological Society* 26:41.

Freuchen, P., and F. Salomonsen. 1958. *The arctic year.* New York: G. P. Putnam's Sons.

Frings, H., and M. Frings. 1957. Recorded calls of the eastern crow as attractants and repellents. *Journal of Wildlife Management* 21:91.

Frings, H., M. Frings, J. Jumber, R. G. Busnel, J. Giban, and P. Gramet. 1958. Reactions of American and French species of *Corvus* and *Larus* to recorded communication signals tested reciprocally. *Ecology* 39:126–131.

Fritz, J., and K. Kotrschal. 1999. Social learning in Common Ravens, *Corvus corax.* *Animal Behaviour* 57:785–793.

Frost, R. 1928. *New Hampshire: A poem with notes and grace notes.* New York: H. Holt.

Garner, K. M. 1978. Management of blackbird and starling winter roost problems in Kentucky and Tennessee. In *Proceedings of the Eighth Vertebrate Pest Conference,* ed. W. E. Howard and R. E. Marsh, 54–59. Davis: Univ. of California Press.

Gibbons, A. 2000. Europeans trace ancestry to Paleolithic people. *Science* 290:1080–1160.

Gilbert, B. 1988. Goodbye, hello. *Sports Illustrated* 69:108–122.

Gillham, O. 2002. *The limitless city: A primer on the urban sprawl debate.* Washington, DC: Island.

Glahn, J. F., E. S. Rasmussen, T. Tomsa, and K. J. Preusser. 1999. Distribution and relative impact of avian predators at aquaculture facilities in the northeastern United States. *North American Journal of Aquaculture* 61:340–348.

Glandt, D., ed. 1991. Der Kolkrabe (*Corvus corax*) in Mitteleuropa. *Metelener Schriftenreihe für Naturschutz* Issue 2, 118 pages.

Glandt, D., and B. Conrad, eds. 2001. Verbreitung und biologie des kolkraben (*Corvus corax*) in Mitteleuropa. *Charadrius* 37:77–136.

Good, E. E. 1952. The life history of the American Crow (*Corvus brachyrhynchos* Brehm). Ph.D. diss., Ohio State Univ., Columbus.

Goodwin, D. 1978. *Birds of man's world.* 2nd ed. London: British Museum.

———. 1986. *Crows of the world.* 2nd ed. London: British Museum.

Gorenzel, W. P., E. L. Fitzhugh, and T. P. Salmon. 1996. Linking GIS to urban crow roost management. *Transactions of the Western Section of the Wildlife Society* 32:48–54.

Gorenzel, W. P., and T. P. Salmon. 1993. Tape-recorded calls disperse American Crows from urban roosts. *Wildlife Society Bulletin* 21:333–338.

———. 1995. Characteristics of American Crow urban roosts in California. *Journal of Wildlife Management* 59:638–645.

Götmark, F., R. Neergaard, and M. Åhlund. 1990. Predation of artificial and real Arctic Loon nests in Sweden. *Journal of Wildlife Management* 54:429–432.

Greene, R. 1592. *Greene's groatsworth of wit: Bought with a million of repentance.* London: William Wright.

Greenwood, P. J. 1980. Mating systems, philopatry and dispersal in birds and mammals. *Animal Behaviour* 28:1140–1162.

Griffin, D. R. 1992. *Animal minds.* Chicago: Univ. of Chicago Press.

Gunther, E. 1945. *Ethnobotany of western Washington: The knowledge and use of indigenous plants by Native Americans.* Seattle: Univ. of Washington Press.

Gurney, J. M. 1921. *Early annals of ornithology.* London: H. F. and G. Witherby.

Ha, R. R., P. Bentzen, J. Marsh, and J. C. Ha. 2003. Kinship and association in social foraging Northwestern Crows (*Corvus corinus*). *Bird Behavior* 15:65–75.

Haase, B. L. 1963. The winter flocking behavior of the Common Crow (*Corvus brachyrynchos brehm*). *Ohio Journal of Science* 63:145–151.

Hall, S. S. 2003. On the trail of West Nile virus. *Smithsonian* 34:88–101.

Hamilton, W. D. 1971. Geometry for the selfish herd. *Journal of Theoretical Biology* 31:295–311.

Hansen, H., C. A. Smedshaug, and G. A. Sonerud. 2000. Preroosting behaviour of Hooded Crows (*Corvus corone cornix*). *Canadian Journal of Zoology* 78:1813–1821.

Hanson, H. G. 1946. Crow center of the United States. *Oklahoma Game and Fish News* 2:4–7.

Hardy, D. E. 1960. *Insects of Hawaii.* Vol. 10. Honolulu: Univ. of Hawaii Press.

Hare, B., M. Brown, C. Williamson, and M. Tomosello. 2002. The domestication of social cognition in dogs. *Science* 298:1634–1636.

Harvey, P. H., and J. R. Krebs. 1990. Comparing brains. *Science* 249:140–146.

Haskell, D., A. M. Knupp, and M. C. Schneider. 2001. Nest predator abundance and urbanization. In *Avian ecology and conservation in an urbanizing world,* ed. J. M. Marzluff, R. Bowman, and R. Donnelly, 245–260. Boston: Kluwer Academic.

Hauser, M. D., and C. Caffrey. 1994. Anti-predator response to raptor calls in wild crows, *Corvus brachyrhynchos hesperis. Animal Behaviour* 48:1469–1471.

Hay, J. 1871. *Pike County ballads and other pieces.* Boston: J. R. Osgood.

Hazelton , E. B. 1969. *Sammy, the crow who remembered.* New York: Charles Scribner's Sons.

Healy, S. D., S. R. de Kort, and N. S. Clayton. 2005. The hippocampus, spatial memory and food hoarding: A puzzle revisited. *Trends in Ecology and Evolution* 20:17–22.

Heinrich, B. 1989. *Ravens in winter.* New York: Summit Books.

———. 1999. *Mind of the raven: Investigations and adventures with wolf-birds.* New York: Cliff Street Books.

Heinrich, B., J. Marzluff, and W. Adams. 1995. Fear and food recognition in native Common Ravens. *Auk* 112:499–503.

Heinrich, B., and J. W. Pepper. 1998. Influence of competitors, on caching behaviour in the Common Raven, *Corvus corax. Animal Behaviour* 56:1083–90.

Henderson, C. R. 1982. An albino American Crow. *Migrant* 53:41.

Heppner, F. 1970. The metabolic significance of differential absorption of radiant energy by black and white birds. *Condor* 72:50–59.

Hickey, J. J. 1943. A guide to bird watching. London: Oxford University Press.

Hill, J., ed. 1999. *An exhilaration of wings: The literature of birdwatching.* New York: Viking Penguin.

Hockett, C. F. 1959. Animal "languages" and human language. In *The evolution of man's capacity for culture,* ed. J. N. Spuhler, 32–39. Chicago: Univ. of Chicago Press.

Hogrefe, T. C., and R. H. Yahner. 1998. Depredation of artificial ground nests in a suburban versus a rural landscape. *Journal of the Pennsylvania Academy of Science* 72:3–6.

Holder, H. 1987. *Crows, an old rhyme.* New York: Farrar, Straus, and Giroux.

Homberg, L. 1957. Fiskande Krakor. *Fauna Och Flora* 5:182–185.

Hope, S. 1980. Call form in relation to function in the Steller's Jay. *American Naturalist* 116:788–820.

———. 1989. Phylogeny of the avian family Corvidae. Ph.D. diss., City Univ. of New York, New York.

Houston, C. S. 1969. Recoveries of the common crow banded in Saskatchewan. *Blue Jay* 27:84–88.

———. 1977. Changing patterns of corvidae on the prairies. *Blue Jay* 35:149–155.

———. 1980. Fall crow roosts in residential Saskatoon. *Blue Jay* 38:42–43.

Howard, H. 1962. *Fossil birds.* Science Series 17. Los Angeles: Los Angeles County Museum.

Huber, B. 1991. Formation, age composition and social structure of nonbreeder groups of the raven (*Corvus corax*). In *Der Kolkrabe (Corvus corax) in Mitteleuropa,* ed. C. D. Glandt, 45–59. Metelener Schriftenreihe für Naturschutz, Issue 2, 118 pages.

Hughes, T. 1971. *Crow.* New York: Harper and Row.

Hunt, G. R. 1996. Manufacture and use of hook-tools by New Caledonian crows. *Nature* 379:249–251.

———. 2000a. Human-like, population-level specialization in the manufacture of pandanus tools by New Caledonian crows (*Corvus moneduloides*). *Proceedings of the Royal Society (London)* 267:403–413.

———. 2000b. Tool use by the New Caledonian Crow *Corvus moneduloides* to obtain Cerambycidae from dead wood. *Emu* 100:109–114.

Hunt, G. R., M. C. Corballis, and R. D. Gray. 2001. Laterality in tool manufacture by crows. *Nature* 414:707.

Hunt, G. R., and R. D. Gray. 2002. Species-wide manufacture of stick-type tools by New Caledonian crows. *Emu* 102:349–353.

Iams, G. 1972. *Fort Cobb crow study.* Oklahoma Department of Wildlife Conservation. Oklahoma City: State project N-82-R-10.

Ignatiuk, J. B. 1991. Breeding biology and habitat selection of American Crows in Saskatchewan parkland habitat. M.Sc. thesis, McGill Univ.

Ignatiuk, J. B., and R. G. Clark. 1991. Breeding biology of American crows in Saskatchewan parkland habitat. *Canadian Journal of Zoology* 69:168–175.

Isack, H. A., and H. U. Reyer. 1989. Honeyguides and honey gatherers: Interspecific communication in a symbiotic relationship. *Science* 243:1343–1346.

James, F. C. 1983. The environmental component of geographic variation in the size and shape of birds: Transplant experiments with blackbirds. *Science* 221:184–186.

James, H. F., and S. L. Olson. 1991. Descriptions of thirty-two new species of birds from the Hawaiian Islands: Part II. Passeriformes. *Ornithological Monographs* 46:3–88.

James, P. C., and N. A. M. Verbeek. 1984. Temporal and energetic aspects of food storage in northwestern crows. *Ardea* 72:207–15.

Jefferson, B. 1991. Evidence of pair bonding between Common Raven (*Corvus corax*) and American Crow (*Corvus brachyrhynchos*). *Ontario Birds* 9:45–48.

———. 1994. Successful hybridization of Common Raven and American Crow. *Ontario Birds* 12:32–35.

Jerzak, L. 2001. Synurbanization of the magpie in the Palearctic. In *Avian ecology and conservation in an urbanizing world,* ed. J. M. Marzluff, R. Bowman, and R. Donnelly, 405–427. Boston: Kluwer Academic.

Johnston, D. W. 1961. *The biosystematics of American Crows*. Seattle: Univ. of Washington Press.

Jones, R. A. 1995. Winged swamp things. *Los Angeles Times* November 1.

Kalmbach, E. R. 1939. *The crow in its relation to agriculture*. Farmers' Bulletin No. 1102. Washington, DC: Department of Agriculture.

Kamil, A. C., and R. P. Balda. 1985. Cache recovery and spatial memory in Clark's Nutcrackers (*Nucifraga columbiana*). *Journal of Experimental Psychology* 11:95–111.

Kaplan, R., and S. Kaplan. 1989. *The experience of nature: A psychological perspective*. Cambridge: Cambridge Univ. Press.

Kelly, J. P., K. L. Etienne, and J. E. Roth. 2002. Abundance and distribution of the Common Raven and American Crow in the San Francisco Bay area, California. *Western Birds* 33:202–17.

Kenward, B., A. A. S. Weir, C. Rutz, and A. Kacelnik. 2005. Tool manufacture by naive juvenile crows. *Nature* 433:121.

Keyes, G. 1998. The name of crow: A cross-cultural survey of terms for the genus Corvus. *Georgia Journal of Anthropology* 2:30–52.

Kilham, L. 1982a. Cleaning/feeding symbioses of common crows with cattle and feral hogs. *Journal of Field Ornithology* 53:275–276.

———. 1982b. Common Crows pulling the tail and stealing food from a River Otter. *Florida Field Naturalist* 10:39–40.

———. 1982c. Florida Red-shouldered Hawk robs American Crows. *Wilson Bulletin* 94:566–567.

———. 1984a. American Crows feeding on and storing River Otter dung. *Florida Field Naturalist* 12:103–104.

———. 1984b. American Crows robbing Great Egrets and White Ibis of large, eel-like salamanders. *Colonial Waterbirds* 7:143–145.

———. 1984c. Cooperative breeding of American Crows. *Journal of Field Ornithology* 55:349–56.

———. 1984d. Foraging and food-storing of American Crows in Florida. *Florida Field Naturalist* 12:25–31.

———. 1984e. Play-like behavior of American Crows. *Florida Field Naturalist* 12:33–36.

———. 1984f. Intra- and extrapair copulatory behavior of American Crows. *Wilson Bulletin* 96:716–717.

———. 1985a. Some breeding season vocalizations of American Crows in Florida. *Florida Field Naturalist* 13:49–76.

———. 1985b. Attacks on fawns, pigs, and other young or weakened mammals by American Crows. *Florida Field Naturalist* 13:17–18.

———. 1985c. Food storing by American Crows in winter. *Connecticut Warbler* 5:8–9.

———. 1985d. Behavior of American Crows in the early part of the breeding cycle. *Florida Field Naturalist* 13:25–48.

———. 1985e. Territorial behavior of American Crows. *Wilson Bulletin* 97:389–390.

———. 1985f. American Crows provoking reactions from Wild Turkeys and Red-shouldered Hawks. *Florida Field Naturalist* 13:94–95.

———. 1986. Vocalizations by female American Crows early in the nesting period. *Journal of Field Ornithology* 57:309–310.

———. 1989. *The American Crow and the Common Raven*. College Station: Texas A&M Univ. Press.

Knight, R. L. 1984. Responses of nesting ravens to people in areas of different human densities. *Condor* 86:345–346.

Knight, R. L., and D. P. Anderson. 1990. Effects of supplemental feeding on an avian scavenging guild. *Wildlife Society Bulletin* 18:388–394.

Knight, R. L., D. P. Anderson, and N. V. Marr. 1991. Responses of an avian scavenging guild to anglers. *Biological Conservation* 56:195–205.

Knight, R. L., D. J. Grout, and S. A. Temple. 1987. Nest-defense behavior of the American Crow in urban and rural areas. *Condor* 89:175–77.

Knight, R. L., H. A. L. Knight, and R. J. Camp. 1993. Raven populations and land-use patterns in the Mojave Desert, California. *Wildlife Society Bulletin* 21:469–71.

———. 1995. Common Ravens and number and type of linear rights-of-way. *Biological Conservation* 74:65–67.

Knopf, F. L., and B. A. Knopf. 1983. Flocking pattern of foraging American Crows in Oklahoma. *Wilson Bulletin* 95:153–55.

Koehler, O. 1950. The ability of birds to count. *Bulletin of the Animal Behaviour Society* 9:41–45.

Koenig, W. D., and F. A. Pitelka. 1979. Relatedness and inbreeding avoidance: Counterploys in the communally nesting Acorn Woodpecker. *Science* 206:1103–1105.

Komar, N., S. Langevin, S. Hinten, N. Nemeth, E. Edwards, D. Hettler, B. Davis, R. Bowen, and M. Bunning. 2003. Experimental infection of North American birds with New York 1999 strain of West Nile virus. *Emerging Infectious Diseases* 9:311–322.

Komar, N., N. A. Panella, J. E. Burns, S. W. Busza, T. M. Mascarenhas, and T. Talbot. 2001. Serologic evidence for West Nile virus infection in birds in the New York City vicinity during an outbreak in 1999. *Emerging Infectious Diseases* 7:621–625.

Konstantinov, V. M. 1996. Anthropogenic transformations of bird communities in the forest zone of the Russian plain. *Acta Ornithologica* 31:53–65.

Krebs, J. R., N. S. Clayton, S. D. Healy, D. A. Criston, S. W. Patel, and A. R. Jolliffe. 1996. The ecology of the brain: Food-storing and the hippocampus. *Ibis* 138:34–46.

Kristan, W. B. III, and W. I. Boarman. 2003. Spatial pattern of risk of Common Raven predation on Desert Tortoises. *Ecology* 84:2432–2443.

Krönneck, P. 1995. Bird remains from Troy, Turkey. In *Archaeozoology of the Near East II*, ed. H. Buitenhuis and H.-P. Uerpmann, 109–112. Leiden: Backhuys.

Kurosawa, R. 1994. Bird abundance in relation to the pavement rate of Tokyo. *Strix* 13:155–164.

———. 1999. An observation of Jungle Crows (*Corvus macrorhynchos*) playing "tennis." *Urban Birds* 16:56–57.

Kurosawa, R., Y. Kanai, and T. Hamaguchi. 2002. Relation of Jungle Crows and garbage in Tokyo III: Comparison between Tokyo and its vicinity and Kawasaki. *Strix* 20:51–59.

Kurosawa, R., M. Narusue, H. Kawachi, and K. Suzuki. 2000. The relationship between Jungle Crows *Corvus macrorhynchos* and garbage in Tokyo. *Strix* 18:71–78.

———. 2001. The relationship between Jungle Crows *Corvus macrorhynchos* and garbage in Tokyo II. *Strix* 19:71–79.

Lachlan, R. F., and M. W. Feldman. 2003. Evolution of cultural communication systems: The coevolution of cultural signals and genes encoding learning preferences. *Journal of Evolutionary Biology* 16:1084–1095.

Laland, K. N., J. Kumm, and M. W. Feldman. 1995. Gene-culture coevolutionary theory: A test case. *Current Anthropology* 36:131–156.

Laland, K. N., J. Odling-Smee, and M. W. Feldman. 2000. Niche construction, biological evolution, and cultural change. *Behavioral and Brain Sciences* 23:131–175.

Lanciotti, R. S., J. T. Roehrig, V. Deubel, J. Smith, M. Parker, K. Steele, B. Crise, K. E. Volpe, M. B. Crabtree, J. H. Scherret, R. A. Hall, J. S. MacKenzie, C. B. Cropp, B. Panigrahy, E. Ostlund, B. Schmitt, M. Malkinson, C. Banet, J. Weissman, N. Komar, H. M. Savage, W. Stone, T. McNamara, and D. J. Gubler. 1999. Origin of the West Nile virus responsible for an outbreak of encephalitis in the northeastern United States. *Science* 286:233–237.

Lariviere, S., and F. Messier. 1998. Effect of density and nearest neighbours on simulated waterfowl nests: Can predators recognize high-density nesting patches? *Oikos* 83:12–20.

Lawrence, J. 1997. *Aesop's fables*. Seattle: Univ. of Washington Press.

Lee, P. 2000. *Norway, the rough guide*. London: Rough Guides.

Legros, D. 1999. *Tommy McGinty's Northern Tutchone story of Crow: First nation elder story of creation*. Gatineau, QE: Canadian Museum of Civilization.

References

Leonard, J. A., R. K. Wayne, J. Wheeler, R. Valadez, S. Guillén, and C. Vilà. 2002. Ancient DNA evidence for Old World origin of New World dogs. *Science* 298:1613-

Lewis, M. L. 1996. Social structure and behavior of wintering American Crows. M.Sc. thesis, Univ. of Alabama, Huntsville.

Löhrl, H. 1968. *Tiere und Wir*. Berlin.

Long, C. A. 1990. American Crow (*Corvus brachyrhynchos*) kills Cooper's Hawk. *Passenger Pigeon* 52:208-209.

Long, C. A., C. Long, and J. E. Long. 1987. Bald Eagle preys on American Crow. *Passenger Pigeon* 49:137.

Lorenz, K. Z. 1952. *King Solomon's ring*. New York: Thomas Y. Crowell.

———. 1981. *The foundations of ethology*. New York: Simon and Schuster.

Luginbuhl, J. M., J. M. Marzluff, and J. E. Bradley. 2001. Corvid survey techniques and the relationship between corvid relative abundance and nest predation. *Journal of Field Ornithology* 72:556-572.

Maccarone, A. D. 1989. Sentinel behaviour in American Crows. *Bird Behaviour* 7:93-95.

———. 1981. Some aspects of feeding and foraging behaviour of three corvids in Newfoundland. M.Sc. thesis, Memorial Univ. of Newfoundland, St. John's.

———. 1991. Seasonal use of foraging habitats by American Crows in central New Jersey. *Bird Behaviour* 9:34-40.

Madge, S., and H. Burn. 1994. *Crows and jays: A guide to the crows, jays, and magpies of the world*. Boston: Houghton Mifflin.

Mainichi Daily News. 2003. Stone the crows! Blackbirds invade baseball game. http://mdn.mainichi.co.jp/news/20030511p2a00m0fp003000c.html.

Malthus, T. R. 1798. *An essay on the principle of population as it affects the future improvement of society, with remarks on the speculations of Mr. Godwin, M. Condorcet and other writers*. London: J. Johnson.

Marr, J. S., and C. H. Calisher. 2003. Alexander the Great and the West Nile virus encephalitis. *Emerging Infectious Diseases* 9:1599-1603.

Martin, L. 1993. *The folklore of birds*. Old Saybrook, CT: Globe Pequot.

Marzluff, J. M. 1988a. Do Pinyon Jays alter nest placement based on prior experience? *Animal Behaviour* 36:1-10.

———. 1988b. Vocal recognition of mates by breeding Pinyon Jays, *Gymnorhinus cyanocephalus*. *Animal Behaviour* 36:296-298.

Marzluff, J. M., and R. P. Balda.1989. Causes and consequences of female-based dispersal in a flock-living bird, the Pinyon Jay. *Ecology* 70:316-328.

———. 1992. *The Pinyon Jay*. London: T. and A. D. Poyser.

References

Marzluff, J. M., R. P. Balda, J. W. Fitzpatrick, and G. E. Woolfenden. 1996. Breeding partnerships of two New World jays. In *Partnerships in birds: The ecology of monogamy,* ed. J. Black, 138–161. Oxford: Oxford Univ. Press.

Marzluff, J. M., R. B. Boone, and G. W. Cox. 1994. Historical changes in populations and perceptions of native pest bird species in the West. *Studies in Avian Biology* 15:202–220.

Marzluff, J. M., and B. Heinrich. 1991. Foraging by Common Ravens in the presence and absence of territory holders: An experimental analysis of social foraging. *Animal Behavior* 42:755–770.

Marzluff, J. M., B. Heinrich, and C. S. Marzluff. 1996. Raven roosts are mobile information centres. *Animal Behaviour* 51:89–103.

Marzluff, J. M., K. J. McGowan, R. Donnelly, and R. L. Knight. 2001. Causes and consequences of expanding American Crow populations. In *Avian ecology and conservation in an urbanizing world,* ed. J. M. Marzluff, R. Bowman, and R. Donnelly, 331–364. Boston: Kluwer Academic.

Mashimo, T., M. Lucas, D. Simon-Chazottes, M.-P. Frenkiel, Z. Montagutelli, P.-E. Ceccaldi, V. Deubel, J.-L. Guénet, and P. Després. 2002. A nonsense mutation in the gene encoding 2'-5'-oligoadenylate synthetase/L1 isoform is associated with West Nile virus susceptibility in laboratory mice. *Proceedings of the National Academy of Sciences (USA)* 99:11311–11316.

Mayr, E. 1974. Behavior programs and evolutionary strategies. *American Scientist* 62:650–659.

McElroy, S. C. 1996. *Animals as teachers and healers.* New York: Ballantine Books.

McGowan, K. J. 2001a. Demographic and behavioral comparisons of suburban and rural American Crows. In *Avian ecology and conservation in an urbanizing world,* ed. J. M. Marzluff, R. Bowman, and R. Donnelly, 365–382. Boston: Kluwer Academic.

———. 2001b. Fish Crow, *Corvus ossifragus. Birds of North America* 589:1–26.

Mesoudi, A., A. Whiten, and K. N. Laland. 2004. Is human cultural evolution Darwinian? Evidence reviewed from the perspective of *The Origin of Species. Evolution* 58:1–11.

Millspaugh, J. J., G. C. Brundige, R. A. Gitzen, and K. J. Radeke. 2000. Elk and hunter space-use sharing in South Dakota. *Journal of Wildlife Management* 64:994–1003.

Mock, D. W., T. C. Lamey, and D. B. A. Thompson. 1988. Falsifiability and the information centre hypothesis. *Ornis Scandinavica* 19:231–248.

Montevecchi, W. A. 1976. Egg size and the egg predatory behaviour of crows. *Behaviour,* 57:3–4.

———. 1978. Corvids using objects to displace gulls from nests. *Condor* 80:349.

Moore, J. E., and P. V. Switzer. 1998. Preroosting aggregations in the American Crow, *Corvus brachyrhynchos. Canadian Journal of Zoology* 76:508–512.

Moore, P. G. 2002. Ravens (*Corvus corax corax* L.) in the British landscape: A thousand years of ecological biogeography in place-names. *Journal of Biogeography* 29:1039–1054.

Morrison, D. W., and D. F. Caccamise. 1990. Comparison of roost use by three species of communal roostmates. *Condor* 92:405–412.

Mueller, H. C. 1968. Prey selection: Oddity or conspicuousness. *Nature* 217:92.

———. 1971. Oddity and specific searching image more important than conspicuousness in prey selection. *Nature* 233:345–346.

Munn, C. A. 1986. Birds that "cry wolf." *Nature* 319:143–145.

Myers, R. 1998. Nothing to crow about. *Seattle Weekly* September 17.

Neatherlin, E. A. 2002. Corvid response to human settlement and recreation with implications for Marbled Murrelets. M.Sc. thesis, Univ. of Washington, Seattle.

Neatherlin, E. A., and J. M. Marzluff. 2004. Campgrounds enable American Crows to colonize remote native forests. *Journal of Wildlife Management* 68:708–718.

Newcombe, C. F. 1923. *Menzies' journal of Vancouver's voyage, April to October, 1792.* Victoria, BC: W. H. Collin.

Nelson, R. K. 1973. *Hunters of the northern forest: Designs for survival among the Alaskan Kutchin.* Chicago: Univ. of Chicago Press.

———. 1983. *Make prayers to the raven: A Koyukon view of the northern forest.* Chicago: Univ. of Chicago Press.

Nemecek, S. 2000. Who were the first Americans? *Scientific American* 283:80–87.

Nicolaus, L. K., and J. F. Cassel. 1983. Taste-aversion conditioning of crows to control predation on eggs. *Science* 220:212–214.

Nihei, Y. Variations of behaviour of Carrion Crows *Corvus corone* using automobiles as nutcrackers. *Japanese Journal of Ornithology* 44:22–28.

Nihei, Y., and H. Higuchi. 2001. When and where did crows learn to use automobiles as nutcrackers? *Tohoku Psychologica Folia* 60:93–97.

Odling-Smee, F. J., K. N. Laland, and M. W. Feldman. 2003. *Niche construction: The neglected process in evolution.* Princeton, NJ: Princeton University Press.

Offbeat News. 2003. "Hungry Hitchcock" ravens kill nineteen sheep. http://www. audarya-fellowship.com/showflat/cat/WorldNews/36947/8/collapsed/5/0/1.

Ogilvie, M. 2003. Albinism, partial albinism and all the other -isms! *Birds of Britain* October. http://www.birdsofbritian.co.uk.

Oliver, J. S. 1994. Estimates of hominid and carnivore involvement in the FLK *Zinjanthropus* fossil assemblage: Some socioecological implications. *Journal of Human Evolution* 27:267–294.

Olson, S. L., and H. F. James. 1991. Descriptions of thirty-two new species of birds from the Hawaiian Islands: Part 1. Non-Passeriformes. *Ornithological Monographs* 45:3–88.

Omland, K. E., C. L. Tarr, W. I. Boarman, J. M. Marzluff, and R. C. Fleischer. 2000. Cryptic genetic variation and paraphyly in ravens. *Proceedings of the Royal Society (London)* 267:2475–2482.

Organ, J. F., and M. R. Ellingwood. 2000. Wildlife stakeholder acceptance capacity for Black Bears, beavers, and other beasts in the East. *Human Dimensions of Wildlife* 5:63–75.

Ortega, J. C., and M. Bekoff. 1987. Avian play: Comparative evolutionary and developmental trends. *Auk* 104:338–341.

Osgood, C. 1937. *The ethnography of the Tanaina.* Yale Univ. Publications in Anthropology No. 16. New Haven: Yale University Press.

———. 1971. *The Hän Indians: A compilation of ethnographic and historical data on the Alaska-Yukon boundary area.* Yale Univ. Publications in Anthropology No. 74. New Haven: Yale University Department of Anthropology.

Paine, R. T. 1966. Food web complexity and species diversity. *American Naturalist* 100:65–75.

Parr, C. S. 1997. Social behavior and long-distance vocal communication in eastern American Crows. Ph.D. diss., Univ. of Michigan, Ann Arbor.

Pavey, C. R., and A. K. Smyth. 1998. Effects of avian mobbing on roost use and diet of powerful owls, *Ninox strenua. Animal Behaviour* 55:331–318.

Paz-y-Miño, G., A. B. Bond, A. C. Kamil, and R. P. Balda. 2004. Pinyon Jays use transitive inference to predict social dominance. *Nature* 430:778–781.

Pearson, R. 1972. *The avian brain.* London: Academic.

Pearson, T. G. 1933. Crows, magpies, and jays. *National Geographic* January:51–79.

Pepperberg, I. M. 1999. *The Alex studies: Cognitive and communicative abilities of Grey Parrots.* Cambridge, MA: Harvard Univ. Press.

Pielou, E. C. 1979. *Biogeography.* New York: John Wiley and Sons.

Pittaway, R. 1988. Wing-tail flicking as a means of distinguishing crows from ravens. *Ontario Birds* 6:74–75.

Philip, N. 1997. *The illustrated book of fairy tales: Spellbinding stories from around the world.* New York: DK.

Pickford, M. 1988. The evolution of intelligence: A palaeontological perspective. In *Intelligence and evolutionary biology,* ed. H. J. Jerison and I. Jerison, 175–198. Berlin: Springer-Verlag.

Pietz, P. J., and D. A. Granfors. 2000. Identifying predators and fates of grassland passerine nests using miniature video cameras. *Journal of Wildlife Management* 64:71–87.

Pimm, S. L. 2001. *The world according to Pimm: A scientist audits the earth.* New York: McGraw-Hill.

Pimm, S. L., M. P. Moulton, and L. J. Justice. 1994. Bird extinctions in the central Pacific. *Philosophical Transactions of the Royal Society* 344:27–33.

Popowski, B. 1962. *The varmint and crow hunter's bible.* New York: Doubleday.

Porter, G. S. 1909. *Birds of the Bible.* Cincinnati: Jennings and Graham.

Portmann, A. 1946. Études sur la cerbralisation chez les oiseaux. I. *Alauda* 14:2–20.

———. 1947. Études sur la cerbralisation chez les oiseaux. II. Les indices intracere-braux. *Alauda* 15:1–15.

Portmann, A., and W. Stingelin. 1961. The central nervous system. In *Biology and comparative physiology of birds,* ed. A. J. Marshall, 1–36. New York: Academic.

Powell, R. W., and W. Kelly. 1977. Tool use in captive crows. *Bulletin of the Psychonomic Society* 10:481–483.

Premack, D. 2004. Is language the key to human intelligence? *Science* 303:318–320.

Price, E. O. 2002. *Animal domestication and behavior.* New York: CABI.

Prior, H., B. Pollok, and O. Güntürkün. 2000. Sich selst vis-à-vis: Was Elstern wahrnehmen. *Rubin* (Ruhr-Universität Bochum) February.

Pryor, K., J. Lindberg, S. Lindbergh, and R. Milano. 1990. A dolphin-human fishing cooperative in Brazil. *Marine Mammal Science* 6:77–82.

Putnam, M. S. 1992. American Crow captures House Sparrow in flight. *Passenger Pigeon* 54:247–249.

Quammen, D. 1985. *Natural acts: A sidelong view of science and nature.* New York: Nick Lyons.

Quiring, D. T., and P. R. Timmins. 1988. Predation by American Crows reduces over-wintering European Corn Borer populations in southwestern Ontario. *Canadian Journal of Zoology* 66:2143–2145.

Raney, P. 1986. Interactions of Fish Crows with American Crows at a landfill near Roswell, Georgia. *Oriole* 51:48.

Ratcliffe, D. 1997. *The raven.* London: T. and A. D. Poyser.

Reaume, T. 1987. Selective feeding by the American Crow. *Ontario Birds* 5:71–72.

———. 1988. Voice of the American Crow. *Ontario Birds* 6:23–24.

Reineke, R. 1995. A comparison of cache retrieval between Gray Jays and American Crows. Ph.D. diss., Univ. of Washington, Seattle.

Rendell, L., and H. Whitehead. 2001a. Culture in whales and dolphins. *Behavioral and Brain Sciences* 24:309–324.

———. 2001b. Cetacean culture: Still afloat after the first naval engagement of the culture wars. *Behavioral and Brain Sciences* 24:360–382.

Renssen, T. A. 1988. Herintroductie van de raaf *Corvus corax* in Nederland. *Limosa* 61:137–144.

Restani, M., and J. M. Marzluff. 2002. Funding extinction? Biological needs and political realities in the allocation of resources to endangered species recovery. *BioScience* 52:169–177.

Restani, M., J. M. Marzluff, and R. E. Yates. 2001. Effects of anthropogenic food sources on movements, survivorship, and sociality of Common Ravens in the Arctic. *Condor* 103:399–404.

Restani, M., R. E. Yates, and J. M. Marzluff. 1996. Capturing Common Ravens *Corvus corax* in Greenland. *Dansk Ornitologisk Forenings Tidsskrift* 90:153–158.

Reuters. 2003. Police pull drunk crow. Reuters News Service October 13.

Richards, D. B., and N. S. Thompson. 1977. Critical properties of the assembly call of the common American Crow. *Behaviour* 64:184–203.

Richards, G. L. 1971. The Common Crow, *Corvus brachyrhynchos,* in the Great Basin. *Condor* 73:116–118.

Richards, G. L., and C. M. White. 1963. Common Crow nesting in Utah. *Condor* 65:530–531.

Richner, H. 1990. Helpers-at-the-nest in Carrion Crows *Corvus corone corone. Ibis* 132:105–108.

———. 1992. The effect of extra food on fitness in breeding Carrion Crows. *Ecology* 73:330–335.

Ridgway, R. 1893. Destruction of crows during the recent cold spell. *Science* 21:77.

Riper, C. van, III, and J. M. Scott. 2001. Limiting factors affecting Hawaiian native birds. *Studies in Avian Biology* 22:221–233.

Robinette, R. L., and C. M. Crockett. 1999. Bald Eagle predation on crows in the Puget Sound region. *Northwestern Naturalist* 80:70–71.

Robinette, R. L., and J. C. Ha. 2000. Beach-foraging behavior of Northwestern Crows as a function of tide height. *Northwestern Naturalist* 81:18–21.

Robinson, L., J. Newell, and J. M. Marzluff. 2005. Twenty-five years of sprawl in the Seattle region: Growth management responses and implications for conservation. *Landscape and Urban Planning* 71:51–72.

Roosevelt, A. C. 2000. Who's on first? *Natural History* 109:76.

Rose, L., and F. Marshall. 1996. Meat eating, hominid sociality, and home bases revisited. *Current Anthropology* 37:307–338.

Røskaft, E. 1985. The effect of enlarged brood size on the future reproductive potential of the rook. *Journal of Animal Ecology* 54:255–260.

Rossow, P. D. 1999. The caching behavior of Common Ravens (*Corvus corax*) at Bluff, Alaska. M.Sc. thesis, Univ. of Alaska, Fairbanks.

Royko, M. 1993. Victim of swooping crows learns that government can't handle every problem. *Columbus Dispatch* June 3:9A.

Sagebiel, J. C. 1998. Pleistocene avian fossils from Zesch cave, Mason County, Texas. *Journal of Vertebrate Paleontology* 18:74A.

Salomonsen, F. 1967. *Grønlands Fugle*.Vol. 3. Copenhagen: Ejnar Munksgaard.

Sarma, V. 1993. *The Pancatantra, book 4: Loss of gains.* London: Penguin Classics.

Sauer, P. 2003. The crows of war: How U.S. foreign policy shapes relations with other species. *Orion* September–October.

Savage, C. 1995. *Bird brains: The intelligence of crows, ravens, magpies, and jays.* Vancouver, BC: Greystone.

Sax, B. 2003. *Crow.* London: Reaktion Books.

Schaefer, J. M. 1983. The Common Crow as a sentinel species of rabies in wildlife populations. Ph.D. diss., Iowa State Univ., Ames.

Schaefer, J. M., and J. J. Dinsmore. 1992. Movement of a nestling between American Crow nests. *Wilson Bulletin* 104:185–187.

Schauer, J. H., and E. C. Murphy. 1996. Predation on eggs and nestlings of Common Murres (*Uria aalge*) at Bluff, Alaska. *Colonial Waterbirds* 19:186–198.

Schultz-Soltau, J. 1962. Rückgang und Wiederausbreitung des Kolkraben (*Corvus corax* L.) im nördlichen Mitteleuropa, unter besonderer Berücksichtigung Niedersachsens. Abhandlungen und Verhandlungen des Naturwissenschaftlichen Vereins in Hamburg, n.s.. 6:337–401.

Scott, J. M., S. Mountainspring, F. L. Ramsey, and C. M. Kepler. 1986. Forest bird communities of the Hawaiian Islands: Their dynamics, ecology, and conservation. *Studies in Avian Biology* 9.

Seattle Times. 2001. Tokyo's had it with crows. September 8.

Septon, F. 1991. American Crow captures adult starling. *Passenger Pigeon* 53:198–199.

Seton, E. T. 1898. *Wild animals I have known.* New York: Charles Scribner's Sons.

Sharpe, R. S., W. R. Silcock, and J. G. Jorgensen. 2001. *Birds of Nebraska: Their distribution and temporal occurrence.* Lincoln: Univ. of Nebraska Press.

Shaw, W. W., and W. R. Mannan. 1984. *Nonconsumptive use of wildlife in the United*

States. U.S. Fish and Wildlife Service Resource Publication, no. 154. Washington, DC: Fish and Wildlife Service.

Shennan, S. 2002. *Genes, memes and human history*. London: Thames and Hudson.

Shida, T. 2001. Traps planned to corral pesky Tokyo crows. *Japan Times* December 15.

Shirota, Y. 1989. A new method to scare crows *Corvus macrorhynchos* and *C. corone*. *Bird Behaviour* 8:1–7.

Simpson, J. A., and E. S. C. Weiner. 1989. *The Oxford English dictionary*. Oxford: Clarendon.

Simmons, K. E. L. 1957. A review of anting-behaviour in birds. *British Birds* 50:401–424.

———. 1966. Anting and the problem of self stimulation. *Journal of Zoology (London)* 149:145–162.

Skagen, S. K., R. L. Knight, and G. H. Orians. 1991. Human disturbance of an avian scavenging guild. *Ecological Applications* 1:215–225.

Skarphédinsson, K. H., Ó. K. Nielsen, S. Thórisson, S. Thorstensen, and S. A. Temple. 1990. *Breeding biology, movements, and persecution of ravens in Iceland*. ACTA Naturalia Islandica. Reykjavik: Icelandic Museum of Natural History.

Slagsvold, T. 1984a. The mobbing behaviour of the Hooded Crow *Corvus corone cornix:* Anti-predator defence of self-advertisement? *Fauna Norvegica Series C. Cinclus* 7:127–131.

———. 1984b. Mobbing behaviour of the Hooded Crow *Corvus corone cornix* in relation to age, sex, size, season, temperature and kind of enemy. *Fauna Norvegica Series C. Cinclus* 8:9–17.

Smedshaug, C. A. 2000. Seasonal variation in area use and roosting behaviour of Hooded Crows, *Corvus corone cornix*. Ph.D. diss., Agricultural Univ. of Norway, Ås.

Smith, A. C. 1887. *The birds of Wiltshire*. London: R. H. Porter and H .F. Bull.

Smith, B. 1998. Fishing with dolphins in the Ayeyarwady River of Myanmar. *Whale and Dolphin Conservation Society Magazine* 19 (online). Available at http://www.wdcs.org.

Smith, E. 1989. Golfers no longer shooting for birdie at Superior course. *Duluth News-Tribune* October 21.

Smith, R. B. 1905. *Bird life and bird lore*. New York: E. P. Dutton.

Soh, M. C. K., N. S. Sodhi, R. K. H. Seoh, and B. W. Brook. 2002. Nest site selection of the House Crow (*Corvus splendens*), an urban invasive bird species in Singapore and implications for its management. *Landscape and Urban Planning* 59:217–226.

Solem, J. K. 1997. Birds observed eating adult Gypsy Moths in June, 1990. *Maryland Birdlife* 53:39.

Sonerud, G. A., C. A. Smedshaug, and Ø. Bråthen. 2001. Ignorant Hooded Crows follow knowledgeable roost-mates to food: Support for the information center hypothesis. *Proceedings of the Royal Society (London)* 268:827–831.

Steenhof, K., M. N. Kochert, and J. A. Roppe. 1993. Nesting by raptors and Common Ravens on electrical transmission line towers. *Journal of Wildlife Management* 57:271–81.

Stein, J. K. 2000. *Exploring Coast Salish prehistory: The archeology of San Juan Island.* Seattle: Univ. of Washington Press.

Stiner, M. C. 2002. Carnivory, coevolution, and the geographic spread of the genus Homo. *Journal of Archaeological Research* 10:1–63.

Stoffel, M. J. 2002. First nesting of Common Ravens in the Saskatoon bird area. *Blue Jay* 60:211–213.

Stouffer, P. C., and D. F. Caccamise. 1991. Roosting and diurnal movements of radio-tagged American Crows. *Wilson Bulletin* 103:387–400.

Struck, D. 2001. Tokyo's winged bullies: City fights losing battle against invading flocks of aggressive crows. *Washington Post* June 12.

Suckley, G., and J. G. Cooper. 1860. *The natural history of Washington Territory and Oregon.* New York: Baillière Brothers.

Sugden, L. G. 1987. Effect of disruptive background on predation of artificial nests by American Crows. *Prairie Naturalist* 19:149–152.

Sullivan, B. D. 1988. Egg predation, home range, and foraging habitat of American Crows in a waterfowl breeding area. M.Sc. thesis, Iowa State Univ., Ames.

——. 1992. Long-eared Owls usurp newly constructed American Crow nests. *Journal of Raptor Research* 26:97–98.

Sullivan, B. D., and J. J. Dinsmore. 1988. Factors affecting egg predation by American Crows. *Journal of Wildlife Management* 54:433–437.

Sutton, G. M. 1951. *Mexican birds.* Norman: Univ. of Oklahoma Press.

Szpir, M. 2003. Why did the crow cross the road? *American Scientist* 91:215–216.

Takagi, K., and K. Ueda. 2002. Distribution of shell-dropping behavior by crows and gulls in Japan. *Strix* 20:61–70.

Tarr, C. L., and R. C. Fleischer. 1999. Population boundaries and genetic diversity in the endangered Mariana Crow (*Corvus kubaryi*). *Molecular Ecology* 8:941–949.

Terres, J. K. 1996. Crow talk. *Birder's World* 10:30–34.

Thompson, F. R. III, and D. E. Burhans. 2003. Predation of songbird nests differs by predator and between field and forest habitats. *Journal of Wildlife Management* 67:407–417.

Thompson, N. S. 1968. Counting and communication in crows. *Communications in Behavioral Biology* A 2:223–225.

———. 1982. A comparison of cawing in the European Carrion Crow (*Corvus corone*) and the American Common Crow (*Corvus brachyrhynchos*). *Behaviour* 80:106–117.

Thompson, P. 1998. Crows may cause abandonment and failure of urban Bald Eagle nests. *WOSNews* 57:6–11.

Thoreau, H. D. 1884. *Walden*. Edinburgh: David Douglass.

Tomiałojc, L. 1979. The impact of predation on urban and rural Woodpigeon (*Columba palumbus* [L.]) populations. *Polish Ecological Studies* 5:141–220.

Tribe of Crow. 1999. *Real Change* 6:8.

Tunnell, G. C. 1996. Reply to Rose and Marshall. *Current Anthropology* 37:327–328.

Twain, M. 1897. *Following the equator: A journal around the world*. Hartford, CT: American Publishing.

Tyler, H. A. 1979. *Pueblo birds and myths*. Univ. of Oklahoma Press.

Tylor, E. B. 1871. *Primitive culture: Researches into the development of mythology, philosophy, religion, art, and custom*. London: J. Murray.

United Nations. 1999. *The state of the world population, 1999: Six billion, a time for choices*. Nafis Sadik, executive director. New York: United Nations.

Valutis, L. L., and J. M. Marzluff. 1999. The appropriateness of puppet-rearing birds for reintroduction. *Conservation Biology* 13:584–591.

Vander Wall, S. B. 1982. An experimental analysis of cache recovery in Clark's Nutcrackers. *Animal Behaviour* 30:84–94.

———. 1990. *Food hoarding in animals*. Chicago: Univ. of Chicago Press.

Vander Wall, S. B., and R. P. Balda. 1981. Ecology and evolution of food-storage behavior in conifer-seed-caching corvids. *Zeitshrift für Tierpsychologie* 56:217–242.

Verbeek, N. A. M., and R. W. Butler. 1981. Cooperative breeding of the Northwestern Crow *Corvus caurinus* in British Columbia. *Ibis* 123:183–189.

Verbeek, N. A. M., and C. Caffrey. 2002. American Crow, *Corvus brachyrhynchos*. *Birds of North America* 647:1–36.

Vigallon, S. M., and J. M. Marzluff. 2005. Is nest predation by Steller's Jays (*Cyanocitta stelleri*) incidental or the result of a specialized search strategy? *Auk* 122:36–49.

Vucetich, J. A., R. O. Peterson, and T. A. Waite. 2004. Raven scavenging favours group foraging in wolves. *Animal Behaviour* 67:1117–1126.

Wallace, A. R. 1881. *Island life*. New York: Harper and Brothers.

Ward, C., and B. S. Low. 1997. Predictors of vigilance for American Crows foraging in an urban environment. *Wilson Bulletin* 109:481–489.

Ward, P., and A. Zahavi. 1973. The importance of certain assemblages of birds as "information-centres" for food-finding. *Ibis* 115:517–534.

Warne, F. L. 1926. Crows is crows. *Bird Lore* 28:110–116.

Watts, J. 2001. Crows with attitude invade Tokyo. *Guardian* April 19.

Weatherhead, P. J. 1983. Two principal strategies in avian communal roosts. *American Naturalist* 121:237–243.

Weir, A. A. S., J. Chappell, and A. Kacelnik. 2002. Shaping of hooks in New Caledonian Crows. *Science* 297:981.

Wells, D. 2002. *One hundred birds and how they got their names*. Chapel Hill, NC: Algonquin Books.

West, B. C., and J. A. Parkhurst. 2002. Interactions between deer damage, deer density, and stakeholder attitudes in Virginia. *Wildlife Society Bulletin* 30:139–147.

White, C. M., and M. Tanner-White. 1988. Use of interstate highway overpasses and billboards for nesting by the Common Raven (*Corvus corax*). *Great Basin Naturalist* 48:64–67.

WhiteBoard News. 1997. Florida man was reunited with his $450 gold bracelet. November 3.

Whitmore, K. D., and J. M. Marzluff. 1998. Hand-rearing corvids for reintroduction: Importance of feeding regime, nestling growth, and dominance. *Journal of Wildlife Management* 62:1460–1479.

Wiles, G. J., J. Bart, R. E. Beck, Jr., and C. F. Aguon. 2003. Impacts of the Brown Tree Snake: Patterns of decline and species persistence in Guam's avifauna. *Conservation Biology* 17:1350–1360.

Wilson, E. O. 1971. *The insect societies*. Cambridge, MA: Harvard Univ. Press.

———. 1984. *Biophilia*. Cambridge, MA: Harvard Univ. Press.

Withey, J. C. 2002. Dispersal behavior of juvenile American Crows and the relationship of crow populations to human population density. M.Sc. thesis, Univ. of Washington, Seattle.

Withey, J. C., and J. M. Marzluff. 2005. Dispersal by juvenile American Crows (*Corvus brachyrhynchos*) influences population dynamics across a gradient of urbanization. *Auk* 122:206–222.

Woodword, T. 1949. *Crow hunting*. Lamar, MO: Journal Publishing.

Woolfenden, G. E., and J. W. Fitzpatrick. 1984. *The Florida Scrub Jay*. Princeton, NJ: Princeton Univ. Press.

Wright, J., R. E. Stone, and N. Brown. 2003. Communal roosts as structured information centers in the raven, *Corvus corax*. *Journal of Animal Ecology* 72:1003–1014.

Wyrost, P. 1993. The fauna of ancient Poland in the light of archaeozoological research. In *Skeletons in her cupboard,* ed. A. Clason, S. Payne, and H.-P. Uerpmann, 251–259. Oxbow Monograph 34. Oxford: Oxbow Books.

Yaremych, S. A. 2003. American Crows and the West Nile virus in east-central Illinois. M.Sc. thesis, Univ. of Illinois, Urbana-Champaign.

Yaremych, S. A., R. E. Warner, P. C. Mankin, J. D. Brawn, A. Raim, and R. Novak. 2004. West Nile virus and high death rate in American Crows. *Emerging Infectious Diseases* 10 (April 2004, available at http://www.cdc.gov/ncidod/EID/vol10no4/03-0499.htm).

Young, L. S., and K. A. Engel. 1985. *Implications of communal roosting by Common Ravens to operation and maintenance of the Malin to Midpoint 500 kV transmission line.* Pacific Power and Light Company Annual Report. Portland, OR: Pacific Power and Light.

Zach, R. 1978. Selection and dropping of whelks by Northwestern Crows. *Behaviour* 67:134–138.

———. 1979. Shell dropping: Decision-making and optimal foraging in Northwestern Crows. *Behaviour* 68:106–117.

Zwickel, F. C., and N. A. M. Verbeek. 1997. Longevity record for the Northwestern Crow, with a comparison to other corvids. *Northwestern Naturalist* 78:11–112.

Index